Selected Titles in This Series

610 **Mark Hovey, John H. Palmieri, and Neil P. Strickland,** Axiomatic stable homotopy theory, 1997

609 **Liviu I. Nicolaescu,** Generalized symplectic geometries and the index of families of elliptic problems, 1997

608 **Christina Q. He and Michel L. Lapidus,** Generalized Minkowski content, spectrum of fractal drums, fractal strings, and the Riemann zeta-functions, 1997

607 **Adele Zucchi,** Operators of class C_0 with spectra in multiply connected regions, 1997

606 **Moshé Flato, Jacques C. H. Simon, and Erik Taflin,** Asymptotic completeness, global existence and the infrared problem for the Maxwell-Dirac equations, 1997

605 **Liangqing Li,** Classification of simple C^*-algebras: Inductive limits of matrix algebras over trees, 1997

604 **Hajnal Andréka, Steven Givant, and István Németi,** Decision problems for equational theories of relation algebras, 1997

603 **Bruce N. Allison, Saeid Azam, Stephen Berman, Yun Gao, and Arturo Pianzola,** Extended affine Lie algebras and their root systems, 1997

602 **Igor Fulman,** Crossed products of von Neumann algebras by equivalence relations and their subalgebras, 1997

601 **Jack E. Graver and Mark E. Watkins,** Locally finite, planar, edge-transitive graphs, 1997

600 **Ambar Sengupta,** Gauge theory on compact surfaces, 1997

599 **Tai-Ping Liu and Yanni Zeng,** Large time behavior of solutions for general quasilinear hyperbolic-parabolic systems of conservation laws, 1997

598 **Valentina Barucci, David E. Dobbs, and Marco Fontana,** Maximality properties in numerical semigroups and applications to one-dimensional analytically irreducible local domains, 1997

597 **Ragnar-Olaf Buchweitz and John J. Millson,** CR-geometry and deformations of isolated singularities, 1997

596 **Paul S. Bourdon and Joel H. Shapiro,** Cyclic phenomena for composition operators, 1997

595 **Eldar Straume,** Compact connected Lie transformation groups on spheres with low cohomogeneity, II, 1997

594 **Solomon Friedberg and Hervé Jacquet,** The fundamental lemma for the Shalika subgroup of $GL(4)$, 1996

593 **Ajit Iqbal Singh,** Completely positive hypergroup actions, 1996

592 **P. Kirk and E. Klassen,** Analytic deformations of the spectrum of a family of Dirac operators on an odd-dimensional manifold with boundary, 1996

591 **Edward Cline, Brian Parshall, and Leonard Scott,** Stratifying endomorphism algebras, 1996

590 **Chris Jantzen,** Degenerate principal series for symplectic and odd-orthogonal groups, 1996

589 **James Damon,** Higher multiplicities and almost free divisors and complete intersections, 1996

588 **Dihua Jiang,** Degree 16 Standard L-function of $GSp(2) \times GSp(2)$, 1996

587 **Stéphane Jaffard and Yves Meyer,** Wavelet methods for pointwise regularity and local oscillations of functions, 1996

586 **Siegfried Echterhoff,** Crossed products with continuous trace, 1996

585 **Gilles Pisier,** The operator Hilbert space OH, complex interpolation and tensor norms, 1996

584 **Wayne W. Barrett, Charles R. Johnson, and Raphael Loewy,** The real positive definite completion problem: Cycle completability, 1996

583 **Jin Nakagawa,** Orders of a quartic field, 1996

582 **Darryl McCollough and Andy Miller,** Symmetric automorphisms of free products, 1996

(Continued in the back of this publication)

Axiomatic Stable Homotopy Theory

MEMOIRS
of the
American Mathematical Society

Number 610

Axiomatic Stable Homotopy Theory

Mark Hovey
John H. Palmieri
Neil P. Strickland

July 1997 • Volume 128 • Number 610 (second of 4 numbers) • ISSN 0065-9266

American Mathematical Society
Providence, Rhode Island

1991 *Mathematics Subject Classification.*
Primary 55P42, 55U35, 55U15, 55N20, 18G35.

Library of Congress Cataloging-in-Publication Data

Hovey, Mark, 1965–
Axiomatic stable homotopy theory / Mark Hovey, John H. Palmieri, Neil P. Strickland.
p. cm. — (Memoirs of the American Mathematical Society, ISSN 0065-9266 ; no. 610)
"July 1997, volume 128, number 610 (second of 4 numbers)."
Includes bibliographical references (p. –) and index.
ISBN 0-8218-0624-6 (alk. paper)
1. Homotopy theory. I. Palmieri, John H. (John Harold), 1964– . II. Strickland, Neil P., 1966– . III. Title. IV. Series.
QA3.A57 no. 610
[QA612.7]
510 s—dc21
[514′.24]

97-12293
CIP

Memoirs of the American Mathematical Society

This journal is devoted entirely to research in pure and applied mathematics.

Subscription information. The 1997 subscription begins with number 595 and consists of six mailings, each containing one or more numbers. Subscription prices for 1997 are \$414 list, \$331 institutional member. A late charge of 10% of the subscription price will be imposed on orders received from nonmembers after January 1 of the subscription year. Subscribers outside the United States and India must pay a postage surcharge of \$30; subscribers in India must pay a postage surcharge of \$43. Expedited delivery to destinations in North America \$35; elsewhere \$110. Each number may be ordered separately; *please specify number* when ordering an individual number. For prices and titles of recently released numbers, see the New Publications sections of the *Notices of the American Mathematical Society.*

Back number information. For back issues see the *AMS Catalog of Publications.*

Subscriptions and orders should be addressed to the American Mathematical Society, P. O. Box 5904, Boston, MA 02206-5904. *All orders must be accompanied by payment.* Other correspondence should be addressed to Box 6248, Providence, RI 02940-6248.

Memoirs of the American Mathematical Society is published bimonthly (each volume consisting usually of more than one number) by the American Mathematical Society at 201 Charles Street, Providence, RI 02904-2294. Periodicals postage paid at Providence, RI. Postmaster: Send address changes to Memoirs, American Mathematical Society, P. O. Box 6248, Providence, RI 02940-6248.

This publication is indexed in *Science Citation Index®*, *SciSearch®*, *Research Alert®*, *CompuMath Citation Index®*, *Current Contents®/Physical, Chemical & Earth Sciences.*
Printed in the United States of America.

∞ The paper used in this book is acid-free and falls within the guidelines established to ensure permanence and durability.
Visit the AMS homepage at URL: `http://www.ams.org/`

10 9 8 7 6 5 4 3 2 1 02 01 00 99 98 97

Contents

ABSTRACT. We define and investigate a class of categories with formal properties similar to those of the homotopy category of spectra. This class includes suitable versions of the derived category of modules over a commutative ring, or of comodules over a commutative Hopf algebra, and is closed under Bousfield localization. We study various notions of smallness, questions about representability of (co)homology functors, and various kinds of localization. We prove theorems analogous to those of Hopkins and Smith about detection of nilpotence and classification of thick subcategories. We define the class of Noetherian stable homotopy categories, and investigate their special properties. Finally, we prove that a number of categories occurring in nature (including those mentioned above) satisfy our axioms.

1991 *Mathematics Subject Classification.* 55P42, 55U35, 55U15, 55N20, 18G35.

Key words and phrases. stable homotopy category, derived category, Bousfield localization, Bousfield class, Brown representability, nilpotence theorem, thick subcategory, Hopf algebra, spectrum, Morava K-theory.

All three authors were partially supported by grants from the National Science Foundation.

1. Introduction and definitions

In algebraic topology, one has many different homology and cohomology theories that one can apply to spaces (such as ordinary cohomology, real or complex K-theory, various kinds of cobordism, and so on). Homotopy theorists realized around 1960 that one could map the category of spaces to the so-called "stable homotopy category," whose objects are known as spectra. In this category all cohomology theories become representable functors, and it is often more convenient to study the representing spectra rather than the cohomology theories themselves. Moreover, many groups of algebraic or geometric interest (such as algebraic K-theory of rings or spaces, or groups of cobordism classes of manifolds) occur as the homotopy groups of spectra.

Of course, cohomology theories occur in many other areas of mathematics. It is our contention that, for many such cohomology theories, there is a stable homotopy category where the cohomology theory in question becomes a representable functor. Just as in algebraic topology, working in this stable homotopy category makes many arguments much simpler and clearer, and suggests new questions. The theory of stable homotopy categories also brings out similarities between different situations.

Our motivation for writing this paper came from three important examples, arising from homological algebra, algebraic topology, and their intersection.

- It is well-known that the derived category of a commutative ring [Ver77, Wei94] has many formal similarities to the stable homotopy category, and Hopkins has proved a nilpotence theorem and a thick subcategory theorem in this algebraic setting (when the ring is Noetherian)—see [Hop87, Nee92a, Tho]. Just how formal are these similarities? That is, is there a common context for both the category of spectra and the derived category so that standard properties of both can be derived simultaneously?
- There are somewhat similar nilpotence theorems for modules over certain Hopf algebras B, including group algebras [QV72] and the Steenrod algebra [Pala]. Over the Steenrod algebra, there are also examples of Bousfield localizations (Margolis' killing construction [Mar83, Pal92]). Is there a good setting—for example, chain complexes of B-modules, or the stable category of [Mar83]—in which to view these results, a setting that would allow one to do "homotopy theory over a Hopf algebra"?
- Fix $n > 0$ and consider the category of $K(n)$-local spectra, much studied by Hopkins and coauthors, for example in [HMS94]. Here $K(n)$ denotes the nth Morava K-theory, one of the most important objects in the ordinary stable homotopy category. How similar is this category to the ordinary category? What sorts of standard homotopy theoretic results hold? Can one represent (co)homology functors or do Bousfield localization? Is there a nilpotence theorem or a thick subcategory theorem?

In this paper, we present a theory of stable homotopy categories to answer these (and many other) questions.

An outline of the paper is as follows. In Section 1.1 we give our axioms for a stable homotopy category. Essentially, a stable homotopy category is a triangulated category (see Appendix A.1) with a smash product and function objects, together with a set of (weak) generators that satisfy Spanier-Whitehead duality. In Section 1.2, we briefly describe some examples, though we postpone all details to Section 9. Section 1.3 has some brief comments about multigrading, needed for

Received by the editor July 18, 1995.

example in the derived category of a graded ring or of a graded Hopf algebra. We close the first section with some definitions and easy lemmas that we use throughout the paper, in Section 1.4.

We begin Section 2 with a discussion, in Section 2.1, of various notions of finiteness in a stable homotopy category. These all coincide in the homotopy category of spectra or in the derived category of a ring, but can be different in general. The most important is the condition that X be small, which means that the functor $[X, -]$ preserves coproducts. In Section 2.2, we consider the (weak) colimits and (weak) limits that exist in a stable homotopy category. In Section 2.3, we combine the previous two sections to construct arbitrary objects from small objects, and to construct certain functors from their restrictions to the subcategory of small objects. Many standard constructions of stable homotopy theory can be found in this section.

Section 3 is about Bousfield localization. This extremely useful construction deserves to be more widely known outside algebraic topology. Section 3.1 contains the basic definitions and properties of localization and colocalization functors; then in Section 3.2, we consider when localization functors exist. Section 3.3 is about various special kinds of localization functors, namely smashing localizations and localizations at or away from a set of small objects. Section 3.4 contains our notion of a morphism of stable homotopy categories, which we call a "geometric morphism." Localization functors provide the main example of such, as we show in Section 3.5. This section also contains the proof that a localization of a stable homotopy category is again a stable homotopy category. In Section 3.6, we consider the notion of Bousfield class; in Section 3.7 we discuss ring objects, field objects, and module objects in a stable homotopy category, and their relation to Bousfield classes. In Section 3.8 we discuss some properties of the collection of smashing localizations.

Section 4 is concerned with the representability of homology functors. This turns out to be a very subtle issue, as the history of Brown representability in the homotopy category of spectra suggests. In Section 4.1 we define the notion of a "Brown category," in which homology functors and maps between them are representable. We also give some conditions that guarantee that a stable homotopy category is a Brown category. In Section 4.2, we discuss some consequences of representability of homology functors. In Section 4.3, we show that a smashing localization of a Brown category is again a Brown category. In Section 4.4, we point out that there is a natural topology on the morphisms in a stable homotopy category.

Section 5 contains a generic approach to nilpotence and thick subcategory theorems analogous to those of Devinatz, Hopkins and Smith [DHS88, HS]. Our theorems are less powerful and more general. Section 5.1 contains our nilpotence theorem, and Section 5.2 contains our thick subcategory theorem.

In Section 6, we discuss the special case of a Noetherian stable homotopy category. A stable homotopy category is Noetherian if the unit of the smash product (i.e., the zero-sphere) is small and the only generator, and if the homotopy of the sphere is a graded Noetherian ring. This case is far simpler than the category of spectra, and includes the derived category of a Noetherian ring and the derived category of comodules over a finite Hopf algebra. In Section 6.1, we show how to split many problems in this context into problems strongly localized at one prime ideal in the homotopy ring of the unit object. In particular, we get can apply our nilpotence theorem to detect nilpotence one prime at a time. In Section 6.2, we apply our thick subcategory theorem. In Section 6.3, we prove an analogue of the telescope conjecture; in certain cases, we can do even better and classify all the localizing subcategories.

Section 7 and Section 8 consider the special cases of connective and semisimple stable homotopy categories, respectively. The connective case, in which the homotopy of the sphere forms a **Z**-graded ring which is zero in negative gradings, is similar to the homotopy category of spectra. The semisimple case is analogous to the category of rational spectra, which is equivalent to the category of graded rational vector spaces, and is thus well-understood.

In Section 9, we consider some examples in more detail. Section 9.1 contains a general approach for constructing stable homotopy categories, commonly known as "cellular approximation." All of our algebraic examples are constructed from chain complexes of some kind, so in Section 9.2, we briefly recall basic facts about chain complexes. Then in Section 9.3, we apply these ideas to show that the derived category of a commutative ring is a stable homotopy category. In Section 9.4 we consider the homotopy category of G-spectra over a complete G-universe $\mathcal{U}$, as in [LMS86] and show how to make it into a stable homotopy category. It is necessary here that $\mathcal{U}$ be complete, although we have a related but unsatisfactory result in the incomplete case. Section 9.5 contains our construction of a stable homotopy category built from comodules over a commutative Hopf algebra. In Section 9.6 we relate this to the commonly studied stable category of modules over a finite cocommutative Hopf algebra, such as a group ring.

Section 10 contains some ideas that we do not as yet fully understand, and some suggestions for future work. For example, Section 10.1 discusses a way to grade a stable homotopy category so that the sphere is the only generator. In the homotopy category of G-spectra, this leads to the theory of Mackey functors. Section 10.2 contains a list of other probable stable homotopy categories that we do not consider in this paper.

We close the paper with two appendices. Appendix A.1 contains a brief discussion of triangulated categories. Triangulated categories are essential for our work, and the reader who has not thought about them in some time would do well to glance at this appendix. Appendix A.2 is a discussion of closed symmetric monoidal categories.

In this paper, we have chosen to work only on the level of homotopy categories. However, we would certainly be interested in a corresponding theory of stable closed model categories [Qui67, DS95]. Ideally there should be a theorem saying that the homotopy category of a closed model category satisfying certain conditions is a stable homotopy category in our sense. There should also be theorems that construct a stable model category from an arbitrary model category, by analogy with the way in which spectra are constructed from spaces. The approach of [Smi] looks particularly promising here. In this connection, we mention the paper [Sch].

We have not attempted to give minimal hypotheses for our theorems, as this would make the paper even more technical than it already is. Many of our results can be proved with much weaker assumptions.

The authors would like to thank many colleagues for their help and support while this paper was being written. These include Hans-Werner Henn, Mike Hopkins, Dan Kan, Chun-Nip Lee, Haynes Miller, Charles Rezk, Hal Sadofsky, and Brooke Shipley. Particular gratitude is due Dan Christensen, who taught us about phantom maps and read many early drafts, and Gaunce Lewis, who helped us considerably with equivariant stable homotopy theory and corrected many of our mistakes.

1.1. **The axioms.** In this section we give our axioms for a stable homotopy category. The central idea is that stable homotopy theory is the study of sufficiently well-behaved triangulated categories.

We shall essentially take as known the concept of a triangulated category, and also the concept of a closed symmetric monoidal category; for details, see Appendices A.1 and A.2 below. The theory of triangulated categories has real content, and familiarity with it is a prerequisite for this paper. The theory of closed symmetric monoidal categories is much more formal—while the definition is complicated, it is an axiomatization of a very familiar and quite simple situation. In a closed symmetric monoidal category, we will usually write $X \wedge Y$ for the monoidal product, S for the unit, and $F(X,Y)$ for the internal function object. In Appendix A.2, we also list a number of natural compatibility requirements between a triangulation and a closed symmetric monoidal structure.

We make three preliminary definitions:

Definition 1.1.1. A *localizing subcategory* of a triangulated category $\mathcal{C}$ is a full subcategory $\mathcal{D}$ such that:

(a) Whenever $X \to Y \to Z \to \Sigma X$ is a cofiber sequence with two of X, Y, and Z objects of $\mathcal{D}$, the third also lies in $\mathcal{D}$.
(b) Any coproduct of objects in $\mathcal{D}$ also lies in $\mathcal{D}$.
(c) If $Y \in \mathcal{D}$ and we have maps $X \xrightarrow{i} Y \xrightarrow{p} X$ with $pi = 1$, then $X \in \mathcal{D}$.

Definition 1.1.2. Let $\mathcal{C}$ be a closed symmetric monoidal additive category, with monoidal product $X \wedge Y$, unit S, and internal function objects $F(X,Y)$. We write $[X,Y]$ for the set of morphisms in $\mathcal{C}$ from X to Y. An object $X \in \mathcal{C}$ is *strongly dualizable* if the natural map $F(X,S) \wedge Y \to F(X,Y)$ is an isomorphism for all Y. An object X is *small* if the natural map $\bigoplus_i [X, Y_i] \to [X, \coprod_i Y_i]$ is an isomorphism, for all coproducts that exist in $\mathcal{C}$.

Definition 1.1.3 (Homology and cohomology functors). Let $\mathcal{C}$ be a triangulated category, and let Ab denote the category of Abelian groups.

(a) An additive functor $H \colon \mathcal{C} \to \mathrm{Ab}$ is *exact* if, whenever
$$X \xrightarrow{f} Y \xrightarrow{g} Z \xrightarrow{h} \Sigma X$$
is a cofiber sequence, the sequence
$$H(X) \xrightarrow{H(f)} H(Y) \xrightarrow{H(g)} H(Z)$$
is exact; and similarly for contravariant additive functors.
(b) An exact functor $H \colon \mathcal{C} \to \mathrm{Ab}$ is a *homology functor* if it takes coproducts to direct sums. An exact contravariant functor $H \colon \mathcal{C}^{\mathrm{op}} \to \mathrm{Ab}$ is a *cohomology functor* if it takes coproducts to products.
(c) Given any object Y of $\mathcal{C}$, we denote the functor $X \mapsto [X,Y]$ by Y^0; then Y^0 is a cohomology functor. Given a cohomology functor H, we say that H is *representable* if there is an object Y of $\mathcal{C}$ and a natural isomorphism of functors between H and Y^0.

There is also a notion of representability for homology functors, analogous to the definition used in ordinary stable homotopy theory but not to the category theorists' definition of representability. Because this is much more subtle than representability of cohomology functors, we defer all definitions and results about it to Section 4.

We can now state the axioms. The following definition was inspired by the axioms in [Mar83, Chapter 2] and the presentation of stable homotopy theory which follows them.

Definition 1.1.4. A *stable homotopy category* is a category $\mathcal{C}$ with the following extra structure:

(a) A triangulation.
(b) A closed symmetric monoidal structure, compatible with the triangulation (as in Definition A.2.1).
(c) A set $\mathcal{G}$ of strongly dualizable objects of $\mathcal{C}$, such that the only localizing subcategory of $\mathcal{C}$ containing $\mathcal{G}$ is $\mathcal{C}$ itself.

We also assume that $\mathcal{C}$ satisfies the following:

(d) Arbitrary coproducts of objects of $\mathcal{C}$ exist.
(e) Every cohomology functor on $\mathcal{C}$ is representable.

We shall say that such a category $\mathcal{C}$ is *algebraic* if the objects of $\mathcal{G}$ are small. If, in addition, the unit object S is small, we say that $\mathcal{C}$ is *unital algebraic*. If, in addition, we have $\mathcal{G} = \{S\}$, we say that $\mathcal{C}$ is *monogenic*. An algebraic stable homotopy category such that all homology functors and natural transformations between them are representable is called a *Brown category* (see Section 4 for the precise definition).

We abuse notation and call the elements of $\mathcal{G}$ *generators*—this is inconsistent with the usual definition in category theory, but is close to the standard meaning in stable homotopy theory. We will also write $\Sigma^*\mathcal{G} = \{\Sigma^n Z \mid Z \in \mathcal{G}, n \in \mathbf{Z}\}$.

In the rest of this paper, $\mathcal{C}$ will denote a stable homotopy category unless otherwise stated.

Not every algebraic stable homotopy category is a Brown category. In fact, not even every monogenic stable homotopy category is a Brown category. See Section 4.1 for details.

Remark 1.1.5.

(a) We were led to this choice of axioms through a long process of generalization. The most familiar stable homotopy categories are monogenic, and even Brown categories. But, as is illustrated in Section 1.2, there are many categories one would like to call stable homotopy categories that are not monogenic or even algebraic. We can as yet prove little about the non-algebraic examples, however.
(b) It is somewhat artificial to insist on a choice of a set $\mathcal{G}$ rather than talking about the thick subcategory generated by $\mathcal{G}$; this is rather like insisting that a manifold come equipped with an atlas rather than an equivalence class of atlases or a maximal atlas. However, the philosophical issues are fairly transparent, and a more philosophically correct approach would entail more verbiage in many places. An enriched triangulated category admits at most one structure as an algebraic stable homotopy category up to a suitable kind of equivalence; see Section 3.4. This is because there is a canonical maximal choice of $\mathcal{G}$, viz. the collection of all small objects (which is essentially a set

by Corollary 2.3.6 and Theorem 2.1.3). However, the $K(n)$-local category can be made into a non-algebraic stable homotopy category (with $\mathcal{G} = \{L_{K(n)}S\}$) or an algebraic one (with $\mathcal{G} = \{L_{K(n)}F\}$, where F is any finite complex of type n).

(c) There are a number of interesting examples of categories satisfying only a subset of our axioms. In particular, there are examples where the generators are not strongly dualizable, but all the other axioms hold. This is the case for the homotopy category of G-spectra indexed on an incomplete universe [LMS86], or sheaves of spectra on a space [BL95]. There are also derived categories of bimodules (where the smash product is not commutative) or of modules over a noncommutative ring (where there is no smash product at all). It would be interesting to work how much of the theory works in these cases (particularly the first). However it turns out that this would increase the complexity of the paper by an unexpectedly large amount, so we have chosen not to address this here.

We shall sometimes want to consider a somewhat weaker notion.

Definition 1.1.6.

(a) An *enriched* additive category is an additive category with arbitrary products and coproducts and an additive closed symmetric monoidal structure. We will sometimes denote the coproduct by $\oplus$ rather than $\amalg$, the symmetric monoidal product by $\otimes$ rather than $\wedge$, the unit by R and the function objects by Hom.

(a) An *enriched* triangulated category is a triangulated category with arbitrary products and coproducts and a compatible closed symmetric monoidal structure in the sense of Appendix A.2.

1.2. **Examples.** In practice, stable homotopy categories seem to arise in two different ways, as indicated by the following two theorems. The following result will be proved as Theorem 2.3.2.

Theorem 1.2.1. *Let $\mathcal{C}$ be an enriched triangulated category, with a given set $\mathcal{G}$ of small, strongly dualizable objects. Suppose also that whenever $[Z, X] = 0$ for all $Z \in \Sigma^*\mathcal{G}$, we have $X = 0$. Then $\mathcal{C}$ is an algebraic stable homotopy category.*

The following result is a subset of Theorems 3.5.1 and 3.5.2; see Section 3 for the definitions.

Theorem 1.2.2. *Suppose that $\mathcal{C}$ is a stable homotopy category, and that $L\colon \mathcal{C} \to \mathcal{C}$ is a localization functor. Then the category $\mathcal{C}_L$ of L-local objects has a natural structure as a stable homotopy category. If $\mathcal{C}$ is algebraic and L is smashing then $\mathcal{C}_L$ is algebraic.*

We (briefly) present a few examples. These are intended to motivate the definitions above, but they can only do that if the reader is familiar with them; so we do not give details of these examples now.

Example 1.2.3. We discuss most of these in detail in Section 9, except for the first one, which we take to be well-known (although we give a few references in Section 9.4).

(a) The homotopy category $\mathcal{S}$ of spectra is a monogenic Brown category: here we have the usual notions of suspension, cofiber sequences (which are the same as fiber sequences, at least up to sign), a smash product, and function spectra. We take $\mathcal{G} = \{S\}$. It is well-known that S is a graded weak generator of $\mathcal{S}$ (in other words, if $[\Sigma^n S, X] = 0$ for all n then $X = 0$) so that Theorem 1.2.1 applies.

(b) Fix a compact Lie group G. The homotopy category $\mathcal{S}_G$ of G-spectra (based on a complete G-universe), is a unital algebraic, non-monogenic Brown category. Once again we have the usual constructions of suspension, cofiber sequences, a smash product, and function spectra. The sphere S is again small. However, it is no longer true that a spectrum X with $[\Sigma^n S, X] = 0$ for all n is zero; we have instead to require that $[\Sigma^n G/H_+, X] = 0$ for all n and all closed subgroups H. We may therefore take

$$\mathcal{G} = \{G/H_+ \mid H \text{ a closed subgroup of } G\}.$$

(c) The (unbounded) derived category $\mathcal{D}(R)$ of a commutative ring R is a monogenic stable homotopy category. An object here can be thought of in many different ways, but all are related to chain complexes of R-modules. So we have a (shift) suspension, cofiber sequences (from the triangulated structure in the homotopy category of chain complexes), a smash product derived from the tensor product of R-modules, and function objects derived from Hom of R-modules. The unit of the smash product is R, and mapping out of R just gives ordinary homology. An object with no homology is trivial. If R is countable, then $\mathcal{D}(R)$ is a Brown category, but this fails in general. Amnon Neeman [Nee95] has shown that if $\mathcal{D}(R)$ is a Brown category, then every flat R-module M has projective dimension at most one, which is false for $R = \mathbf{C}[x, y]$ and $M = \mathbf{C}(x, y)$.

(d) Let B be a commutative Hopf algebra over a field k. Let $\mathcal{C}(B)$ be the category whose objects are (unbounded) cochain complexes of injective B-comodules, with morphisms given by cochain homotopy classes of maps. Then $\mathcal{C}(B)$ is a unital algebraic stable homotopy category. We have a (shift) suspension, cofiber sequences and a smash product derived from the tensor product (over k) of B-comodules. A slightly less well-known construction gives function objects. The unit S of the smash product is an injective resolution of k, and S is small. As in the G-equivariant stable homotopy category, S is not a graded weak generator; we have to take $\mathcal{G}$ to be the set of injective resolutions of simple comodules. Note that if I and J are injective resolutions of comodules M and N, respectively, then $[I, J]_* \simeq \operatorname{Ext}^*_B(M, N)$. The category $\mathcal{C}(B)$ is monogenic if k is the only simple comodule (say, if $B = (kG)^*$ for G a p-group, k a field of characteristic p). Note that if B is graded connected, then $\mathcal{C}(B)$ will not be monogenic, because the set of simple comodules is the set of all of the internal suspensions of k. It will be monogenic in the multigraded sense defined in Section 1.3. In general, $\mathcal{C}(B)$ will probably not be a Brown category.

(e) Let B be a finite-dimensional commutative Hopf algebra over a field k. A B-comodule is just the same thing as a module over the cocommutative Hopf algebra B^*. We can construct the stable comodule category $\operatorname{StComod}(B)$ from the category of B-comodules by killing maps that factor through injective comodules. This can also be described as $\operatorname{StMod}(B^*)$, of course, in which we kill off maps that factor through projectives. It is a unital algebraic stable homotopy category, and if k is countable then it is a Brown category.

The suspension functor takes M to the cokernel of any inclusion of M into an injective comodule. Because B is finite, it turns out that this functor is an equivalence. The cofiber sequences are derived from exact sequences, the smash product from the tensor product (over k), and the function objects from Hom. The unit of the smash product is k, and k is small. In the case $B = (kG)^*$, the homotopy groups $\pi_* M = [k, M]_*$ are the Tate cohomology groups of M. Just as in part (d), there are nontrivial objects with no homotopy. We take the set $\mathcal{G}$ in StComod(B) to be the set of non-projective simple comodules. Note that $[M, N]_s = \mathrm{Ext}_B^s(M, N)$ if $s > 0$. The category StComod(B) is monogenic if k is the only simple comodule.

(f) Let E be a commutative S-algebra in the sense of [EKMM95] (this is morally the same as an E_∞ ring spectrum in older foundational settings). It follows easily from the results of [EKMM95] that the derived category of E-modules is a monogenic stable homotopy category. The smash products and function objects are analogous to the tensor product and Hom over a ring. The unit is E itself, and the resulting homotopy groups $\pi_* X = [E, X]_*$ of X agree with the homotopy groups of the underlying spectrum. In particular, E is small, and an object with no homotopy is trivial. However, this category will not be a Brown category in general. Similar things can be done equivariantly, in which case the resulting category will be unital algebraic but not monogenic.

(g) In the homotopy category of spectra $\mathcal{S}$, one often considers Bousfield localization functors $L\colon \mathcal{S} \to \mathcal{S}$ (Definition 3.1.1). Let $\mathcal{S}_L$ denote the category of local spectra, i.e., spectra X that are equivalent to LY for some Y. As stated above in Theorem 1.2.2, $\mathcal{S}_L$ is a stable homotopy category. Consider in particular the localization functor $L = L_{K(n)}$ with respect to Morava K-theory. In this case, $\mathcal{S}_L$ is algebraic, and in fact a Brown category, but not unital. If F is a finite spectrum of type n then LF is small and we can take $\mathcal{G} = \{LF\}$, but LS is not small. This is our primary example of an algebraic, non-unital stable homotopy category, and it is studied more closely in [HSS]. However, there are many analogous examples. One could take the localization of $\mathcal{D}(\mathbf{Z})$ with respect to $\mathbf{F}_p$, for example.

1.3. **Multigrading.** In some parts of this paper, we will allow our categories to be multigraded. In other words, we shall assume that (for some $d \geq 1$) we have a family of "spheres" S^k for each $k \in \mathbf{Z}^d$, with coherent isomorphisms $S^k \wedge S^l \simeq S^{k+l}$ and $S^{(j,0,\ldots,0)} = \Sigma^j S$. We shall also assume the usual sign rule, that the composite

$$S^{k+l} \simeq S^k \wedge S^l \xrightarrow{\text{twist}} S^l \wedge S^k \simeq S^{k+l}$$

is just multiplication by $(-1)^m$, where $m = \sum_i k_i l_i$. Given $k \in \mathbf{Z}^d$, we shall write $\Sigma^k X = S^k \wedge X$. In this context, if we say that $f\colon X \to Y$ is a graded morphism, we mean that $f\colon \Sigma^k X \to Y$ for some $k \in \mathbf{Z}^d$.

If $\mathcal{C}$ is a $\mathbf{Z}^d$-graded unital algebraic stable homotopy category with $\mathcal{G} = \{S^k \mid k \in \mathbf{Z}^d\}$, then we say that $\mathcal{C}$ is *monogenic* in the multigraded sense.

1.4. **Some basic definitions and results.** In this section, we give some basic definitions we will use throughout the paper. We also draw some basic consequences of our axioms. These include the fact that the generators detect isomorphisms (Lemma 1.4.5) and the existence of arbitrary products (Lemma 1.4.7).

For most of this section, $\mathcal{C}$ will be a stable homotopy category, though many of the definitions and results will hold in greater generality.

For the definition of homology and cohomology functors, see Definition 1.1.3. We have defined homology and cohomology functors to be ungraded, but of course we will sometimes need to consider gradings.

Definition 1.4.1 (Grading).

(a) If H is a homology functor on $\mathcal{C}$, we make H into a graded functor by defining $H_nX = H(\Sigma^{-n}X)$. If H is a cohomology functor, we define $H^nX = H(\Sigma^{-n}X)$. If H and K are both homology functors and $\tau\colon H \longrightarrow K$ is a natural transformation, then τ induces a natural transformation $\tau_*\colon H_* \longrightarrow K_*$ compatible with the suspension. We have a similar remark for cohomology functors.
(b) Similarly, we define the graded Abelian group $[X,Y]_*$ by $[X,Y]_n = [\Sigma^n X, Y]$.
(c) We write $S^k = \Sigma^k S$ and $\pi_k(X) = [S^k, X]$. We refer to $\pi_k(X)$ as the kth *homotopy group* of X. This is not a homology functor unless S is small.

Definition 1.4.2. A category $\mathcal{D}$ is *essentially small* if it has only a set of isomorphism classes. Note that this has nothing to do with the objects of $\mathcal{D}$ being small in the sense of Definition 1.1.2.

Definition 1.4.3 (Types of subcategories of $\mathcal{C}$). Let $\mathcal{C}$ be a stable homotopy category.

(a) A subcategory $\mathcal{D}$ of $\mathcal{C}$ is called *thick* if $\mathcal{D}$ is closed under cofibrations and retracts. That is, if there is an exact triangle $X \longrightarrow Y \longrightarrow Z \longrightarrow \Sigma X$ with two of X, Y, Z in $\mathcal{D}$, then so is the third; and if we have $Y \in \mathcal{D}$ and maps $X \xrightarrow{i} Y \xrightarrow{p} X$ with $pi = 1$ then $X \in \mathcal{D}$. Thus, if $X \amalg Z \in \mathcal{D}$ then $X, Z \in \mathcal{D}$.
(b) A subcategory $\mathcal{D}$ is a *localizing* subcategory if it is thick and closed under arbitrary coproducts. That is, if we have a set of objects $\{X_i\}$ of $\mathcal{D}$, then $\coprod X_i$ is in $\mathcal{D}$.
(c) Dually, a subcategory $\mathcal{D}$ is a *colocalizing* subcategory if it is thick and closed under arbitrary products ($\mathcal{C}$ has arbitrary products, by Lemma 1.4.7).
(d) A thick subcategory $\mathcal{D}$ is an *ideal* if $Y \wedge X \in \mathcal{D}$ for all $Y \in \mathcal{C}$ whenever $X \in \mathcal{D}$. We call $\mathcal{D}$ a *localizing ideal* if it is both a localizing subcategory and an ideal.
(e) A thick subcategory $\mathcal{D}$ is a *coideal* if $F(Y,X) \in \mathcal{D}$ for all $Y \in \mathcal{C}$ whenever $X \in \mathcal{D}$. We call $\mathcal{D}$ a *colocalizing coideal* if it is both a colocalizing subcategory and a coideal.
(f) A thick subcategory $\mathcal{D}$ is a $\mathcal{G}$-*ideal* if $Y \wedge X \in \mathcal{D}$ whenever $X \in \mathcal{D}$ and $Y \in \mathcal{G}$.
(g) A thick subcategory $\mathcal{D}$ is a $\mathcal{G}$-*coideal* if $F(Y,X) \in \mathcal{D}$ whenever $X \in \mathcal{D}$ and $Y \in \mathcal{G}$.

The reason for the names "localizing" and "colocalizing" will become clear in Section 3.

We point out that there is a real difference between localizing subcategories and localizing ideals. If G is a compact Lie group and $\mathcal{C}$ is the category of G-spectra, then $\mathcal{D} = \{X \in \mathcal{C} \mid X^G = 0\}$ is a localizing subcategory but not a localizing ideal (because $G_+ \in \mathcal{D}$ but $F(G_+, G_+) = F(G_+, S) \wedge G_+ \notin \mathcal{D}$). However in the monogenic case, every localizing subcategory is a localizing ideal.

For any collection $\mathcal{S}$ of objects of a triangulated category $\mathcal{C}$, the intersection $\mathcal{D}$ of all thick subcategories containing $\mathcal{S}$ is itself a thick subcategory, and it is clearly the smallest thick subcategory that contains $\mathcal{S}$. We refer to $\mathcal{D}$ as the thick subcategory *generated* by $\mathcal{S}$. Similar comments apply to the other types of subcategories considered above. We write

$$\begin{aligned}
\operatorname{loc}\langle \mathcal{S}\rangle &= \text{localizing subcategory generated by } \mathcal{S}\\
\operatorname{coloc}\langle \mathcal{S}\rangle &= \text{colocalizing subcategory generated by } \mathcal{S}\\
\operatorname{locid}\langle \mathcal{S}\rangle &= \text{localizing ideal generated by } \mathcal{S}\\
\operatorname{colocid}\langle \mathcal{S}\rangle &= \text{colocalizing coideal generated by } \mathcal{S}\\
\operatorname{thick}\langle \mathcal{S}\rangle &= \text{thick subcategory generated by } \mathcal{S}
\end{aligned}$$

The last of these is characterized more constructively in the proof of Proposition 2.3.5. Similar characterizations could be given using transfinite recursion in the other cases, but this does not seem very useful.

Remark 1.4.4. For any objects X and Y in a stable homotopy category $\mathcal{C}$, it is clear that $\{Z \mid Z \wedge Y \in \operatorname{loc}\langle X \wedge Y\rangle\}$ is a localizing subcategory containing X, and thus that $\operatorname{loc}\langle X\rangle \wedge Y \subseteq \operatorname{loc}\langle X \wedge Y\rangle$.

Lemma 1.4.5. *Let $\mathcal{C}$ be a stable homotopy category.*

(a) *Suppose that $\tau\colon H(W) \longrightarrow K(W)$ is a morphism of homology or cohomology functors, which is an isomorphism whenever $W \in \Sigma^*\mathcal{G}$. Then τ is an isomorphism for all $W \in \mathcal{C}$.*
(b) *Suppose that $X \in \mathcal{C}$ is such that $[W, X] = 0$ for all $W \in \Sigma^*\mathcal{G}$. Then $X = 0$.*
(c) *Suppose that $f\colon Y \longrightarrow Z$ is a map that induces an isomorphism $[W, Y] \longrightarrow [W, Z]$ for every $W \in \Sigma^*\mathcal{G}$. Then f is an isomorphism.*

Proof. (a): Write

$$\mathcal{D} = \{W \mid \tau_*\colon H_*W \longrightarrow K_*W \text{ is an isomorphism }\}.$$

This is a localizing subcategory, and contains $\mathcal{G}$, so it is all of $\mathcal{C}$.

(b): By a similar argument, $[W, X] = 0$ for all $W \in \mathcal{C}$. In particular, $[X, X] = 0$, so $X = 0$.

(c): Apply (b) to the fiber of f. □

Lemma 1.4.6. *Let $\mathcal{C}$ be a stable homotopy category. Then any localizing $\mathcal{G}$-ideal in $\mathcal{C}$ is an ideal, and any colocalizing $\mathcal{G}$-coideal is a coideal. If $\mathcal{G} = \{S\}$ (in particular, if $\mathcal{C}$ is monogenic) then every (co)localizing subcategory is a (co)ideal.*

Proof. Suppose that $\mathcal{D}$ is a localizing $\mathcal{G}$-ideal. Write

$$\mathcal{E} = \{W \mid W \wedge X \in \mathcal{D} \text{ for all } X \in \mathcal{D}\}.$$

Then $\mathcal{E}$ is a localizing subcategory containing $\mathcal{G}$, so it is the whole of $\mathcal{C}$. Similarly, if $\mathcal{D}$ is a colocalizing $\mathcal{G}$-coideal, then let

$$\mathcal{E} = \{W \mid F(W, X) \in \mathcal{D} \text{ for all } X \in \mathcal{D}\}.$$

Then $\mathcal{E}$ is a localizing subcategory containing $\mathcal{G}$, so is all of $\mathcal{C}$. The last sentence follows trivially. □

We shall say that a triangulated category is *complete* if it has arbitrary products, and *cocomplete* if it has arbitrary coproducts. Thus, a stable homotopy category is by definition cocomplete. Note that this is inconsistent with the usual definition in category theory: (co)completeness usually means that all (co)limits exist. Only in rather uninteresting cases do triangulated categories have equalizers or coequalizers. We will see in Propositions 2.2.4 and 2.2.11 that they do have "weak" limits and colimits.

Lemma 1.4.7. *Every stable homotopy category is complete.*

Proof. Given a family $\{Y_i\}$ of objects, we need to construct their product. To do so, we represent the cohomology functor that takes X to $\prod[X, Y_i]$. □

Another useful fact is that idempotents in a stable homotopy category always split.

Lemma 1.4.8. *Let X be an object in a stable homotopy category $\mathcal{C}$. Suppose that $e \in [X, X]$ is idempotent (that is, $e^2 = e$). Then there is an isomorphism $f\colon X \longrightarrow Y \amalg Z$ for some $Y, Z \in \mathcal{C}$, such that the following diagram commutes.*

$$\begin{array}{ccccc} X & & \xrightarrow{\quad e \quad} & & X \\ f\Big\downarrow\simeq & & & & \simeq\Big\downarrow f \\ Y \amalg Z & \longrightarrow & Y & \longrightarrow & Y \amalg Z \end{array}$$

(The maps on the bottom row are the obvious ones.)

Proof. Observe that $X^0U = [U, X]$ is a cohomology functor, which splits as a direct sum of two cohomology functors

$$[U, X] \simeq e[U, X] \oplus (1 - e)[U, X].$$

By the definition of a stable homotopy category, the two cohomology functors on the right are representable, say $e[U, X] \simeq [U, Y]$ and $(1 - e)[U, X] \simeq [U, Z]$. We immediately get a diagram as above with X, Y and Z replaced by the functors X^0, Y^0 and Z^0, to which we can apply the Yoneda lemma. □

This has the following entertaining corollary:

Lemma 1.4.9 (The Eilenberg swindle)**.** *Let $\mathcal{C}$ be a stable homotopy category, and $\mathcal{D}$ a subcategory closed under coproducts and cofibrations. Then $\mathcal{D}$ is also closed under retracts, and thus is a localizing subcategory.*

Proof. Suppose that $W \in \mathcal{D}$ and we have maps $X \xrightarrow{i} W \xrightarrow{p} X$ with $pi = 1$. Then $ip\colon W \longrightarrow W$ is an idempotent, and the lemma shows that there is a splitting $W = X \amalg Y$ for some Y. Write

$$Z = X \amalg (Y \amalg X) \amalg (Y \amalg X) \amalg \cdots = (X \amalg Y) \amalg (X \amalg Y) \amalg \ldots.$$

From the second description, we see that $Z \in \mathcal{D}$. There is an evident isomorphism $Z \simeq X \amalg Z$, so $X \amalg Z \in \mathcal{D}$. The cofiber of the inclusion map $Z \longrightarrow X \amalg Z$ is just X, so $X \in \mathcal{D}$. □

Because triangulated categories are additive, finite coproducts and finite products coincide. In an algebraic stable homotopy category, many infinite coproducts and products coincide as well:

Proposition 1.4.10. *In an algebraic stable homotopy category $\mathcal{C}$, the natural map $\coprod Y_i \longrightarrow \prod Y_i$ is an equivalence if and only if, for all $Z \in \Sigma^*\mathcal{G}$, we have $[Z, Y_i] = 0$ for all but finitely many i.*

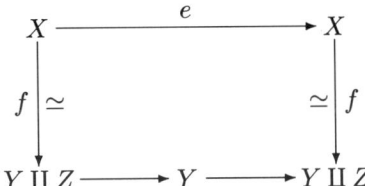

Proof. Since the generators are small, the map

$$[Z, \coprod Y_i] = \bigoplus [Z, Y_i] \to \prod [Z, Y_i] = [Z, \prod Y_i]$$

will be an equivalence if and only if the condition of the proposition is satisfied. Lemma 1.4.5 completes the proof. □

Remark 1.4.11. It is possible to use representability of cohomology functors and the smash product to construct function objects; indeed, one just represents the cohomology functor $Z \mapsto [Z \wedge X, Y]$ to get the function object $F(X, Y)$. This will make any symmetric monoidal triangulated category such that cohomology functors are representable into a closed symmetric monoidal category, but this structure may not be compatible with the triangulation (though we do not have a counterexample). The difficulty comes in proving that $F(X, Y)$ is an exact functor of X.

2. Smallness, limits and constructibility

In this section, we address three related questions. Firstly, we examine various different notions of smallness or finiteness for an object X in a stable homotopy category. Secondly, we consider different kinds of limits and colimits that might exist in such a category. Finally, we study various possible senses in which an object X can be constructed from a family $\mathcal{A}$ of small objects.

2.1. **Notions of finiteness.**

We collect here various definitions that restrict the size of an object Z in a stable homotopy category. Some of these have been given before, but we repeat them for ease of reference.

Definition 2.1.1 (Notions of finiteness)**.** Consider an object Z of $\mathcal{C}$. We say that Z is:

(a) *small* if for any collection of objects $\{X_i\}$, the natural map $\bigoplus [Z, X_i] \to [Z, \coprod X_i]$ is an isomorphism.
(b) F*-small* if for any collection of objects $\{X_i\}$, the natural map $\coprod F(Z, X_i) \to F(Z, \coprod X_i)$ is an isomorphism.
(c) $\mathcal{S}$*-finite* (for any family $\mathcal{S}$ of objects of $\mathcal{C}$) if Z lies in the thick subcategory $\operatorname{thick}\langle \mathcal{S} \rangle$ generated by $\mathcal{S}$.
(d) *strongly dualizable* if for any X, the natural map $DZ \wedge X \to F(Z, X)$ is an equivalence.

Remark 2.1.2. The most important case of part (c) is the case $\mathcal{S} = \mathcal{G}$, but other cases do arise. In the $K(n)$-local category, for instance, it is most natural to take $\mathcal{G} = \{Z\}$, where $Z = L_{K(n)} Z'$ for some finite type n spectrum Z', while it is more natural to take "finite" to mean $\{S\}$-finite rather than $\{Z\}$-finite.

In any case, in this paper we often use $\mathcal{F}$ (or $\mathcal{F}_{\mathcal{C}}$) to denote the full subcategory of small objects of $\mathcal{C}$; by Theorem 2.1.3 below, if $\mathcal{C}$ is algebraic, then $\mathcal{F} = \operatorname{thick}\langle \mathcal{G} \rangle$.

Theorem 2.1.3. *Let $\mathcal{C}$ be a stable homotopy category.*

(a) *Suppose that X is small (respectively F-small, or strongly dualizable) and Y is strongly dualizable. Then $X \wedge Y$ is also small (or F-small, or strongly dualizable). Thus, the categories of small, F-small, and strongly dualizable objects are all $\mathcal{G}$-ideals.*

We also have the following implications.

(b) *$\mathcal{G}$-finite $\Rightarrow$ strongly dualizable $\Leftrightarrow$ F-small.*
(c) *If $\mathcal{C}$ is algebraic, then small $\Leftrightarrow$ $\mathcal{G}$-finite $\Rightarrow$ strongly dualizable $\Leftrightarrow$ F-small.*
(d) *If $\mathcal{C}$ is unital algebraic, then small $\Leftrightarrow$ $\mathcal{G}$-finite $\Leftrightarrow$ strongly dualizable $\Leftrightarrow$ F-small.*
(e) *If $\mathcal{C}$ is algebraic, any $\mathcal{G}$-ideal of small objects is closed under the Spanier-Whitehead duality functor D.*

Proof. (a): Suppose that X is small and Y is strongly dualizable. We then have

$$\begin{aligned}[X \wedge Y, \coprod Z_i] &= [X, DY \wedge \coprod Z_i] \\ &= [X, \coprod DY \wedge Z_i] \\ &= \bigoplus [X, DY \wedge Z_i] \\ &= \bigoplus [X \wedge Y, Z_i].\end{aligned}$$

Thus $X \wedge Y$ is small. The proof when X is F-small is similar. Now suppose instead that X is strongly dualizable. We find that

$$\begin{aligned}F(X \wedge Y, Z) &= F(X, F(Y, Z)) \\ &= DX \wedge DY \wedge Z \\ &= F(X, DY) \wedge Z \\ &= D(X \wedge Y) \wedge Z.\end{aligned}$$

Thus, $X \wedge Y$ is strongly dualizable. It is easy to see that the categories of small, F-small and strongly dualizable objects are all thick; as they are closed under smashing with a strongly dualizable object, they are in fact $\mathcal{G}$-ideals.

(b): The category of strongly dualizable objects is thick and contains $\mathcal{G}$, so it contains all $\mathcal{G}$-finite objects. If X is strongly dualizable then

$$F(X, \coprod Y_i) = DX \wedge \coprod Y_i = \coprod DX \wedge Y_i = \coprod F(X, Y_i).$$

Thus X is F-small.

Conversely, suppose that X is F-small, and let $\mathcal{D}$ be the category of those Y for which the natural map $DX \wedge Y \to F(X, Y)$ is an isomorphism. Using part (e) of Theorem A.2.5, we see that every strongly dualizable object lies in $\mathcal{D}$, in particular $\mathcal{G} \subset \mathcal{D}$. Moreover, it is easy to see that $\mathcal{D}$ is localizing, so $\mathcal{D} = \mathcal{C}$. Thus X is strongly dualizable.

(c): First note that $\mathcal{G}$ consists of small objects because $\mathcal{C}$ is algebraic, so every $\mathcal{G}$-finite object is small. The converse will be proved as Corollary 2.3.12 below. Thus small $\Leftrightarrow$ $\mathcal{G}$-finite. The other claims in (c) are covered by (b).

(d): Suppose that $X \in \mathcal{C}$ is F-small; it is enough to show that X is small. This follows immediately by applying the homology functor $[S, -]$ to the equality $F(X, \coprod Y_i) = \coprod F(X, Y_i)$.

(e): Now suppose that $\mathcal{C}$ is algebraic, $\mathcal{D}$ is a $\mathcal{G}$-ideal of small objects, and $X \in \mathcal{D}$. Then X is strongly dualizable, so by Lemma A.2.6, DX is a retract of $DX \wedge X \wedge DX$. In particular, using part (a), we find that DX is small. This means DX is $\mathcal{G}$-finite, by part (c). Since $\mathcal{D}$ is a $\mathcal{G}$-ideal, we conclude that $DX \wedge X \wedge DX \in \mathcal{D}$, so $DX \in \mathcal{D}$. □

Note that this theorem gives evidence that the choice of generators in an algebraic stable homotopy category is not very relevant, although we shall not attempt to make this precise here.

One might ask whether every $\mathcal{G}$-finite object lies in the triangulated category generated by $\mathcal{G}$, or whether one really needs to use retractions as well as cofibers. Retractions are necessary when $\mathcal{C}$ is the derived category of the ring $\mathbf{Q} \times \mathbf{Q}$ (a connective, semisimple, monogenic Brown category). However, suppose that $\mathcal{C}$ is a connective stable homotopy category, and write $R = \pi_0 S$. If R is Noetherian and has finite global dimension, and finitely generated projective R-modules are free, then every $\mathcal{G}$-finite object lies in the triangulated category generated by $\mathcal{G}$ (see Section 7).

2.2. **Weak colimits and limits.** In this section, we discuss colimits and limits in a stable homotopy category. Almost never will actual colimits and limits exist, but weak versions always exist.

Definition 2.2.1. Let $\mathcal{C}$ be a triangulated category. Fix a small category $\mathcal{I}$; we will write i for a typical object. Given an object X of $\mathcal{C}$, let c_X denote the functor $\mathcal{I} \longrightarrow \mathcal{C}$ which is constant at X. Given a map $g\colon X \longrightarrow Y$ in $\mathcal{C}$, let $c_g\colon c_X \longrightarrow c_Y$ denote the obvious natural transformation.

Given any functor $i \mapsto X_i$ from $\mathcal{I}$ to $\mathcal{C}$, we say that a pair (X, τ) is a *weak colimit* of (X_i) if

1. X is an object of $\mathcal{C}$.
2. τ is a natural transformation from (X_i) to c_X.
3. Given any natural transformation $\rho\colon (X_i) \longrightarrow c_Y$, there is a map $g\colon X \longrightarrow Y$ so that $\rho = c_g \tau$. Equivalently, the natural map
$$[X, Y] \longrightarrow \varprojlim{}_i [X_i, Y]$$
is surjective for all Y.

Note that the map g need not be unique. The pair (X, τ) will be called a *minimal weak colimit* if the map $\varinjlim{}_i HX_i \longrightarrow HX$ induced by τ is an isomorphism for all homology functors $H\colon \mathcal{C} \longrightarrow \mathrm{Ab}$. (This definition is mainly useful when $\mathcal{C}$ is algebraic.)

Our definition of minimal weak colimit is not the obvious generalization of the one given in [Mar83, Chapter 3]. However, our definition is often equivalent to that of Margolis. Indeed, we have the following proposition, whose proof we will defer to Proposition 4.2.1.

Proposition 2.2.2. *Let $\mathcal{C}$ be a Brown category. Suppose that $\mathcal{I}$ is a small category, $i \mapsto X_i$ is a functor from $\mathcal{I}$ to $\mathcal{C}$, and $(\tau_i\colon X_i \longrightarrow X)$ is a weak colimit. Then X is the minimal weak colimit if and only if the induced map*
$$\varinjlim [Z, X_i]_* \longrightarrow [Z, X]_*$$
is an isomorphism for all $Z \in \mathcal{G}$.

The definition could be modified in various ways when $\mathcal{C}$ is not a Brown category. At present we know few examples in that context; our present definition handles them better than any of the variants, but that could easily change if more examples come to light.

The question of existence of a minimal weak colimit for a given diagram is subtle (except when the diagram is a sequence). Even if the diagram can be rigidified in some underlying closed model category, there is no reason in general that the homotopy colimit should be a weak colimit in the homotopy category. See Theorem 4.2.3 for a result in this direction.

We also make the following more constructive definition.

Definition 2.2.3. Given a sequence

$$X_0 \xrightarrow{f_0} X_1 \xrightarrow{f_1} X_2 \xrightarrow{f_2} \dots$$

in a stable homotopy category $\mathcal{C}$, define the *sequential colimit* to be the cofiber of the map

$$F\colon \coprod X_i \to \coprod X_i$$

that takes the summand X_i to $X_i \amalg X_{i+1}$ by $1_{X_i} - f_i$. This is often called the *telescope* of the X_i, but we prefer a more consistent terminology for the various different kinds of colimits.

In particular, given a self-map $f\colon \Sigma^d X \to X$ of an object X, we write $f^{-1}X$ for the sequential colimit of the sequence

$$X \xrightarrow{f} \Sigma^{-d}X \xrightarrow{f} \Sigma^{-2d}X \dots.$$

We have the following proposition.

Proposition 2.2.4. *Let $\mathcal{I}$ be a small category and $\mathcal{C}$ a stable homotopy category.*

(a) *Every functor $i \mapsto X_i$ from $\mathcal{I}$ to $\mathcal{C}$ has a weak colimit.*

(b) *Suppose that $(\tau_i\colon X_i \to X)$ and $(\sigma_i\colon Y_i \to Y)$ are weak colimits, and that $(u_i\colon X_i \to Y_i)$ is a natural transformation. Then there is a compatible map $u\colon X \to Y$ (typically not unique) such that the following diagram commutes:*

$$\begin{array}{ccc} X_i & \xrightarrow{\tau_i} & X \\ {\scriptstyle u_i}\downarrow & & \downarrow{\scriptstyle u} \\ Y_i & \xrightarrow{\sigma_i} & Y \end{array}$$

(c) *Suppose that $\mathcal{C}$ is algebraic and that $(X_i \to X)$ is a minimal weak colimit. Then any compatible map (as in (b)) from X to any other weak colimit X' is a split monomorphism, so X is a retract of X'. If X' is also minimal, then X is non-canonically isomorphic to X'.*

(d) *The sequential colimit of a sequence is a minimal weak colimit. In fact, for any Y there is a Milnor exact sequence*

$$0 \to \varprojlim{}^1_i[\Sigma X_i, Y] \to [X, Y] \to \varprojlim{}_i[X_i, Y] \to 0.$$

(e) *If $Y \in \mathcal{C}$ and $(\tau_i\colon X_i \to X)$ is a (minimal) weak colimit then $(\tau_i \wedge 1\colon X_i \wedge Y \to X \wedge Y)$ is a (minimal) weak colimit.*

(f) *Suppose that $\mathcal{C}$ is algebraic and $\mathcal{D} \subseteq \mathcal{C}$ is a localizing subcategory. If $(X_i \to X)$ is a minimal weak colimit in $\mathcal{C}$ with each $X_i \in \mathcal{D}$, then $X \in \mathcal{D}$.*

Proof. The proof of most of this is the same as for the analogous propositions in [Mar83, Chapter 3]. That is, to construct a weak colimit of (X_i), we consider the cofiber of the map

$$\coprod_{\alpha \in \operatorname{mor} \mathcal{I}} X_{\operatorname{dom}(\alpha)} \xrightarrow{F} \coprod_{i \in \operatorname{ob} \mathcal{I}} X_i$$

where $X_{\mathrm{dom}(\alpha)}$ maps to $X_{\mathrm{dom}(\alpha)}$ by the identity and to $X_{\mathrm{codom}(\alpha)}$ by $-\alpha$. It is easy to verify that this is a weak colimit (proving (a)), but it is almost never minimal. Part (b) follows easily.

Suppose we have a compatible map $u\colon X \longrightarrow X'$ as in part (c). By (b), we also have a compatible map $v\colon X' \longrightarrow X$. Thus $vu\colon X \longrightarrow X$ is compatible with the identity map of (X_i). It follows that for any $Z \in \Sigma^*\mathcal{G}$, the map vu induces the identity on $[Z, X] = \varinjlim [Z, X_i]$. By Lemma 1.4.5 we see that vu is an isomorphism, so u is a split monomorphism. If X' is also minimal then uv is an isomorphism by the same argument, so u and v are isomorphisms. This proves (c).

Next, consider a sequence $(X_i : i \in \mathbf{N})$ as in (d), and write X for the sequential colimit. By applying $[-, Y]$ to the cofibration which defines X, we obtain a long exact sequence

$$[\coprod_i X_i, Y] \xleftarrow{F^*} [\coprod_i X_i, Y] \leftarrow [X, Y] \leftarrow [\coprod_i \Sigma X_i, Y] \xleftarrow{F^*} [\coprod_i \Sigma X_i, Y].$$

From this we extract a short exact sequence $A_{*+1} \longrightarrow [X, Y]_* \longrightarrow B_*$, where A_* and B_* are the cokernel and kernel of the map $\prod_i [X_i, Y]_* \longrightarrow \prod_i [X_i, Y]_*$. These are by definition just $\lim^1_i [X_i, Y]_*$ and $\lim_i [X_i, Y]_*$, so we get a Milnor exact sequence as stated in (d). For the right hand map to be surjective means precisely that X is a weak colimit of the X_i. Now suppose that H is a homology functor, so that $H(\coprod X_i) = \bigoplus_i HX_i$. One can check directly that the induced map $F_*\colon H(\coprod X_i) \longrightarrow H(\coprod X_i)$ is injective, with cokernel $\varinjlim_i H(X_i)$. It follows easily that X is the *minimal* weak colimit. This proves (d).

Suppose that $(\tau_i\colon X_i \longrightarrow X)$ is a weak colimit. We claim that $(\tau_i \wedge 1\colon X_i \wedge Y \longrightarrow X \wedge Y)$ is also a weak colimit. To see this, suppose we have compatible maps $X_i \wedge Y \longrightarrow Z$. By adjunction we get maps $X_i \longrightarrow F(Y, Z)$; as X is a weak colimit of the X_i, we get a map $X \longrightarrow F(Y, Z)$; by adjunction we get a map $X \wedge Y \longrightarrow Z$. It is easy to check that this has the required property. Suppose moreover that X is the minimal weak colimit, and that H is a homology functor. Then $H(- \wedge Y)$ is also a homology functor, so $H(X \wedge Y) = \varinjlim_i H(X_i \wedge Y)$. It follows that $X \wedge Y$ is the minimal weak colimit of the objects $X_i \wedge Y$. This proves (e).

By the proof of (a), if each object in a diagram is in a localizing subcategory, then that diagram has a weak colimit that is also in the localizing subcategory. A minimal weak colimit is a retract of any other weak colimit, so it will also be in the localizing subcategory. This proves (f). □

Remark 2.2.5. As observed by Boardman (see [Bou83]), the result of part (f) is true for homotopy colimits. Suppose that $\mathcal{C}$ is a stable homotopy category that arises from a suitable closed simplicial model category. Let $\{X_i\}$ be a diagram in this underlying category, such that each object X_i lies in a given localizing subcategory $\mathcal{D} \subseteq \mathcal{C}$. The claim is that the homotopy colimit (X, say) also lies in $\mathcal{D}$. Indeed, one can show that X is homotopy equivalent to the sequential colimit of a sequence $X(0) \longrightarrow X(1) \longrightarrow \dots$ of cofibrations, such that $X(k)/X(k-1)$ is a coproduct of suspensions of X_i's.

We pause to prove two useful facts about sequential colimits.

Lemma 2.2.6. *If $f\colon \Sigma^d X \longrightarrow X$ is an isomorphism, then there is a natural isomorphism $X \longrightarrow f^{-1}X$.*

Proof. We will omit the suspensions from this proof. Write $Y = \coprod_{i=0}^{\infty} X$, and let X_i be the ith copy of X inside Y. Recall that $f^{-1}X$ is the cofiber of the map $F\colon Y \to Y$ that takes the summand X_i to $X_i \amalg X_{i+1}$ by $(1, -f)$. Let $J\colon X \to Y$ be the inclusion of X_0, and let $Q\colon Y \to X$ be the map that is $f^{-k}\colon X \to X$ on X_k. Finally, let $G\colon Y \to Y$ be the map that sends X_k to $X_0 \amalg \ldots \amalg X_{k-1}$ by $(-f^{-k}, \ldots, -f^{-1})$. One can then check that $QJ = 1$, $QF = 0$, $GJ = 0$, $GF = 1$ and $FG + JQ = 1$. Thus F is a split monomorphism, and J identifies X with the cokernel (or cofiber) of F. □

Let $\mathbf{N}$ denote the natural numbers, $\mathbf{N} = \{0, 1, 2, \ldots\}$.

Lemma 2.2.7. *Let $\{X_k\}_{k \in \mathbf{N}}$ be a directed system of objects of a stable homotopy category $\mathcal{C}$. Suppose that $u\colon \mathbf{N} \to \mathbf{N}$ is a weakly increasing map, such that $u(k) \to \infty$ as $k \to \infty$. Then we have an isomorphism of sequential colimits*

$$\varinjlim{}_k X_{u(k)} \xrightarrow{\simeq} \varinjlim{}_k X_k$$

Proof. We define two maps

$$G, H\colon \coprod_k X_{u(k)} \to \coprod_k X_k$$

as follows. The map G just sends the kth summand $X_{u(k)}$ in the source to the $u(k)$th summand in the target by the identity. The map H sends $X_{u(k)}$ to $X_{u(k)} \amalg \ldots \amalg X_{u(k+1)-1}$; the component $X_{u(k)} \to X_m$ is just the map provided by the direct system. We also write F for the usual map, whose cofiber is by definition the sequential colimit. It is straightforward to check that we get a commutative square as on the left of the following diagram, and thus a map

$$f\colon \varinjlim{}_k X_{u(k)} \to \varinjlim{}_k X_k$$

as indicated.

$$\begin{array}{ccccccc}
\coprod_k X_{u(k)} & \xrightarrow{F} & \coprod_k X_{u(k)} & \longrightarrow & \varinjlim{}_k X_{u(k)} & \longrightarrow & \Sigma \coprod_k X_{u(k)} \\
{\scriptstyle H}\big\downarrow & & \big\downarrow{\scriptstyle G} & & \big\downarrow{\scriptstyle f} & & \big\downarrow{\scriptstyle \Sigma H} \\
\coprod_k X_k & \xrightarrow{F} & \coprod_k X_k & \longrightarrow & \varinjlim{}_k X_k & \longrightarrow & \Sigma \coprod_k X_k
\end{array}$$

Consider the resulting Milnor exact sequences for $[\varinjlim{}_k X_k, Y]$ and $[\varinjlim{}_k X_{u(k)}, Y]$. It is well-known that the relevant $\varprojlim$ and $\varprojlim^1$ terms are isomorphic, so that

$$[\varinjlim{}_k X_k, Y] = [\varinjlim{}_k X_{u(k)}, Y].$$

As this holds for all Y, Yoneda's lemma tells us that f is an isomorphism. □

Remark 2.2.8. One might ask whether the sequential colimit of cofiber sequences is a cofiber sequence. Provided that there is a suitable underlying closed model category, it turns out that this is true, in the following weak sense. Suppose that we have cofiber sequences $X_k \xrightarrow{u_k} Y_k \xrightarrow{v_k} Z_k \xrightarrow{w_k} \Sigma X_k$ and commutative diagrams

$$\begin{array}{ccc}
X_k & \xrightarrow{u_k} & Y_k \\
{\scriptstyle f_k}\big\downarrow & & \big\downarrow{\scriptstyle g_k} \\
X_{k+1} & \xrightarrow{u_{k+1}} & Y_{k+1}
\end{array}$$

Then it is possible to choose maps $h_k: Z_k \longrightarrow Z_{k+1}$, and compatible maps

$$X_\infty \xrightarrow{u_\infty} Y_\infty \xrightarrow{v_\infty} Z_\infty \xrightarrow{w_\infty} \Sigma X_\infty$$

of the sequential colimits, such that (f_k, g_k, h_k) is a morphism of triangles, and $(u_\infty, v_\infty, w_\infty)$ is a cofiber sequence.

Indeed, we can make a telescope construction to replace the sequence $\{X_k\}$ by a weakly equivalent sequence of cofibrations (with X_k cofibrant). We can then inductively modify the Y's, u's and g's to get a weakly equivalent diagram in which the g's and u's are cofibrations and the $X-Y$ squares commute on the nose. Having done this, the claim is fairly clear.

We do not know whether this is true in an arbitrary stable homotopy category.

We also have the notion of a weak limit, dual to that of a weak colimit.

Definition 2.2.9. Let $\mathcal{I}$ be a small category. Given any functor $i \mapsto X_i$ from $\mathcal{I}$ to $\mathcal{C}$, we say that a pair (X, τ) is a *weak limit* of (X_i) if

1. X is an object of $\mathcal{C}$.
2. τ is a natural transformation from c_X to (X_i).
3. Given any natural transformation $\rho: c_Y \longrightarrow (X_i)$, there is a map $g: Y \longrightarrow X$ so that $\rho = \tau c_g$. Equivalently, the natural map

$$[Y, X] \longrightarrow \varprojlim{}_i [Y, X_i]$$

is surjective for all Y.

Note that the map g need not be unique.

We also have the simpler definition of a sequential limit. Recall that stable homotopy categories always have arbitrary products, by Lemma 1.4.7.

Definition 2.2.10. Let $\mathcal{C}$ be a stable homotopy category Given a sequence of objects of $\mathcal{C}$,

$$X_0 \xleftarrow{f_0} X_1 \xleftarrow{f_1} X_2 \xleftarrow{f_2} \dots$$

define the *sequential limit* to be the fiber of the map

$$F: \prod X_i \leftarrow \prod X_i$$

such that $\pi_i \circ F = \pi_i - f_i \circ \pi_{i+1}$.

We do not know a good notion of a minimal weak limit in any of our settings. However, we do have the following two propositions, whose proofs are analogous to that of Proposition 2.2.4. We leave the proofs to the interested reader.

Proposition 2.2.11. *Let $\mathcal{I}$ be a small category, and $\mathcal{C}$ a stable homotopy category.*

(a) *Every functor $i \mapsto X_i$ from $\mathcal{I}$ to $\mathcal{C}$ has a weak limit.*

(b) *Suppose that $(\tau_i: X \longrightarrow X_i)$ and $(\sigma_i: Y \longrightarrow Y_i)$ are weak limits, and that $(u_i: X_i \longrightarrow Y_i)$ is a natural transformation. Then there is a compatible map $u: X \longrightarrow Y$ (typically not unique).*

(c) *The sequential limit of a sequence is a weak limit. In fact, for any Y there is a Milnor exact sequence*

$$0 \longrightarrow \varprojlim{}^1 [Y, \Sigma^{-1} X_i] \longrightarrow [Y, X] \longrightarrow \varprojlim [Y, X_i] \longrightarrow 0.$$

(d) *If* $Y \in \mathcal{C}$ *and* $(\tau_i \colon X \to X_i)$ *is a weak (resp., sequential) limit then the diagram*

$$(F(Y,\tau_i) \colon F(Y,X) \to F(Y,X_i))$$

is a weak (resp., sequential) limit.

Proposition 2.2.12. *Suppose that* $(\tau_i \colon X_i \to X)$ *is a weak (resp., sequential) colimit in a stable homotopy category* $\mathcal{C}$*, and* Y *is an object of* $\mathcal{C}$*; then the diagram*

$$(F(\tau_i,Y) \colon F(X,Y) \to F(X_i,Y))$$

is a weak (resp., sequential) limit. □

Sequential limits are exact in a limited sense, analogous to Remark 2.2.8.

2.3. **Cellular towers and constructibility.** In this section, we consider the problem of constructing an object from a given family of objects $\mathcal{A}$. We first consider the case $\mathcal{A} = \Sigma^*\mathcal{G}$.

Proposition 2.3.1. *Suppose that* $\mathcal{C}$ *is an algebraic stable homotopy category. Then every object* X *can be written as the sequential colimit of a sequence* $0 = X^0 \to X^1 \to \dots$*, in which the cofiber of each map* $X^k \to X^{k+1}$ *is a coproduct of objects of* $\Sigma^*\mathcal{G}$.

Proof. Suppose that $X \in \mathcal{C}$. Let $X_0 = X$, and let S_0 be the coproduct

$$\coprod_{Z \in \Sigma^*\mathcal{G}} \coprod_{f \in [Z,X_0]} Z.$$

There is an obvious map $S_0 \to X_0$ which induces a surjection $[Z,S_0] \to [Z,X_0]$ for all $Z \in \Sigma^*\mathcal{G}$. Let X_1 be the cofiber of this map. By iterating this construction, we get cofibrations $S_k \to X_k \to X_{k+1}$ in which S_k is a coproduct of copies of objects in $\Sigma^*\mathcal{G}$, and the map $[Z,X_k] \to [Z,X_{k+1}]$ is zero for every $Z \in \Sigma^*\mathcal{G}$.

Now let X^k denote the fiber of the map $X \to X_k$. Using the octahedral axiom, we get a diagram as follows:

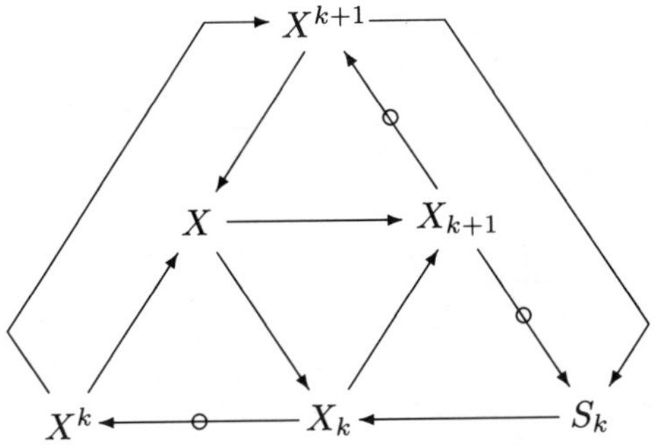

In particular, we get a sequence of maps $0 = X^0 \to X^1 \to \dots \to X$, in which the cofiber of $X^{k-1} \to X^k$ is S_k. Let CX be the sequential colimit. By the weak colimit property, we get a map $CX \to X$ compatible with the given maps $X^k \to X$.

Suppose that $Z \in \Sigma^*\mathcal{G}$. By construction, the map $[Z,X] \to [Z,X_k]$ is zero for $k > 0$, so the map $[Z,X^k] \to [Z,X]$ is surjective. Moreover, the map $[\Sigma Z, S_k] \to$

$[\Sigma Z, X_k]$ is surjective; after a diagram chase, we conclude that the kernel of the map $[Z, X^k] \to [Z, X]$ goes to zero in $[Z, X^{k+1}]$. It follows easily that $[Z, CX] = \varinjlim_k [Z, X^k] = [Z, X]$, and thus (by lemma 1.4.5) that the map $CX \to X$ is an isomorphism. $\square$

This construction is called the *cellular tower* for X.

We now restate and prove Theorem 1.2.1.

Theorem 2.3.2. *Let $\mathcal{C}$ be an enriched triangulated category. Suppose that $\mathcal{G}$ is a set of small strongly dualizable objects of $\mathcal{C}$. Suppose also that whenever $[Z, X] = 0$ for all $Z \in \Sigma^*\mathcal{G}$, we have $X = 0$. Then $\mathcal{C}$ is an algebraic stable homotopy category.*

Proof. We need to show both that the only localizing subcategory of $\mathcal{C}$ that contains $\mathcal{G}$ is $\mathcal{C}$ itself, and that every cohomology functor is representable. Suppose that $X \in \mathcal{C}$. As in the proof of Proposition 2.3.1, we construct a sequence of objects $0 = X^0 \to X^1 \to X^2 \to \dots$. We define CX to be the sequential colimit, and obtain a map $CX \to X$, with cofiber LX, say. Just as above, we find that this induces an isomorphism $[Z, CX] \to [Z, X]$ for all $Z \in \Sigma^*\mathcal{G}$, so that $[Z, LX]_* = 0$ for all such Z. Thus $LX = 0$ and $X \simeq CX$. By construction, CX lies in the localizing subcategory $\mathrm{loc}\langle \mathcal{G} \rangle$ generated by $\mathcal{G}$.

Now let H be a cohomology functor on $\mathcal{C}$. We need to show that H is representable. This is much the same as [Mar83, Theorem 4.11]. We shall define recursively a sequence of objects

$$X(0) \xrightarrow{i_0} X(1) \xrightarrow{i_1} X(2) \xrightarrow{i_2} \dots$$

and elements $u(k) \in H(X(k))$ such that $i_k^* u(k+1) = u(k)$. We start with

$$X(0) = \coprod_{Z \in \Sigma^*\mathcal{G}} \coprod_{v \in H(Z)} Z.$$

We take $u(0)$ to be the element of

$$H(X(0)) = \prod_{Z \in \Sigma^*\mathcal{G}} \prod_{v \in H(Z)} H(Z)$$

whose (Z, v)th component is v. We then set

$$T(k) = \{(Z, f) \mid Z \in \Sigma^*\mathcal{G},\ f \colon Z \to X(k),\ f^* u(k) = 0\}.$$

We define $X(k+1)$ by the cofiber sequence

$$\coprod_{(Z,f) \in T(k)} Z \to X(k) \xrightarrow{i_k} X(k+1).$$

By applying H to this, we obtain a three-term exact sequence (with arrows reversed). It is clear by construction that $u(k)$ maps to zero in the left hand term, so that there exists $u(k+1) \in H(X(k+1))$ with $i_k^* u(k+1) = u(k)$ as required.

We now let X be the sequential colimit of the objects $X(k)$. The cofibration defining this sequential colimit gives rise to a short exact sequence

$$0 \to \varprojlim{}^1_k H(\Sigma X(k)) \to H(X) \to \varprojlim_k H(X(k)) \to 0.$$

Using this, we find an element $u \in H(X)$ that maps to $u(k)$ in each $H(X(k))$. As in Yoneda's lemma, this induces a natural map $\tau_U \colon [U, X] \to H(U)$. It is easy to see that τ_Z is an isomorphism for each $Z \in \Sigma^* \mathcal{G}$ (using the fact that these objects are small). It is also easy to see that

$$\{Z \mid \tau_{\Sigma^k Z} \text{ is an isomorphism for all } k\}$$

is a localizing category. It contains $\mathcal{G}$, so it must be all of $\mathcal{C}$; thus τ is an isomorphism. □

We now prove some technical results about the cardinality of various categories. Recall that for infinite cardinals κ and λ we have

$$\kappa + \lambda = \kappa\lambda = \max(\kappa, \lambda).$$

Definition 2.3.3. Given $X \in \mathcal{C}$, we define a cardinal number $c(X)$ by

$$c(X) = \sum_{Z \in \Sigma^* \mathcal{G}} |[Z, X]|.$$

Note that $c(X) > \max(|\mathcal{G}|, \aleph_0)$. We also define

$$\mathcal{C}_\kappa = \{X \in \mathcal{C} \mid c(X) \leq \kappa\}$$

and

$$c(\mathcal{C}) = \sum_{Z \in \mathcal{G}} c(Z) = \sum_{Z, W \in \mathcal{G}, n \in \mathbf{Z}} |[W, Z]_n|.$$

It is not hard to see that $\mathcal{C}_\kappa$ is a thick subcategory. Moreover, if $\{X_i\}_{i \in I}$ is a family of objects in $\mathcal{C}_\kappa$ and $|I| \leq \kappa$, then $\coprod_i X_i \in \mathcal{C}_\kappa$. Finally, if $\kappa \geq \max(|\mathcal{G}|, \aleph_0)$ then

$$\mathcal{C}_\kappa = \{X \mid |[Z, X]| \leq \kappa \text{ for all } Z \in \Sigma^* \mathcal{G}\}.$$

Proposition 2.3.4. *Suppose that $\mathcal{C}$ is an algebraic stable homotopy category, and that $\kappa \geq c(\mathcal{C})$. Then*

(a) *$X \in \mathcal{C}_\kappa$ if and only if X is the sequential colimit of a cellular tower*

$$0 = X^0 \longrightarrow X^1 \longrightarrow \dots$$

such that the cofiber of $X^i \longrightarrow X^{i+1}$ is a coproduct of suspended generators indexed by a set of cardinality at most κ.

(b) *$\mathcal{C}_\kappa$ is essentially small. That is, the isomorphism classes of objects of $\mathcal{C}_\kappa$ form a set.*

Proof. Certainly if X is such a sequential colimit, then $|[Z, X]| \leq \kappa$ for all $Z \in \Sigma^* \mathcal{G}$, since the generators are small. Conversely, if $X \in \mathcal{C}_\kappa$, then in the construction of the cellular tower as in Proposition 2.3.1, we only need to take coproducts over sets of cardinality at most κ. These cellular towers give us explicit models for isomorphism classes of objects of $\mathcal{C}_\kappa$ so we can use them to construct a small skeleton. More precisely, let $\mathcal{A}_0$ denote the set of all coproducts of suspensions of generators indexed by sets of size at most κ. Having defined $\mathcal{A}_n$, for each map from an object of $\mathcal{A}_0$ to an object of $\mathcal{A}_n$, choose a cofiber. Denote the set of such choices by $\mathcal{A}_{n+1}$. Let $\mathcal{A}_\infty$ be the union of the $\mathcal{A}_n$, and for each sequence $X_1 \longrightarrow X_2 \longrightarrow \dots$ in $\mathcal{A}_\infty$, choose a sequential colimit. Let $\mathcal{A}$ denote the set of such choices. Then the cellular tower constructed above shows that any object of $\mathcal{C}_\kappa$ is isomorphic to an element of $\mathcal{A}$. □

Proposition 2.3.5. *Let $\mathcal{C}$ be a stable homotopy category, and $\mathcal{S}$ a set of objects of $\mathcal{C}$. Then the thick subcategory generated by $\mathcal{S}$ is essentially small, as is the $\mathcal{G}$-ideal generated by $\mathcal{S}$.*

Proof. Without loss of generality, we may assume that $\mathcal{S}$ is closed under suspensions and desuspensions. We shall recursively define sets $\mathcal{S}_k \subseteq \mathcal{C}$ for each integer $k \geq 0$, closed under suspensions and desuspensions, starting with $\mathcal{S}_0 = \mathcal{S}$. Suppose that we have constructed $\mathcal{S}_k$. Every retract of an object $X \in \mathcal{S}_k$ corresponds to an idempotent $e \in [X, X]$, so there are only a set of these, up to isomorphism; choose one in each isomorphism class. Similarly, there are only a set of maps $f\colon X \longrightarrow Y$ with $X, Y \in \mathcal{S}_k$, and thus only a set of cofibers, up to isomorphism; choose one in each isomorphism class. Let $\mathcal{S}_{k+1}$ be the union of $\mathcal{S}_k$ with the set of all these choices, so that $\mathcal{S}_{k+1}$ is again a set, closed under suspensions and desuspensions. It is easy to see that the set $\bigcup_k \mathcal{S}_k$ is equivalent to the thick subcategory generated by $\mathcal{S}$.

The proof for $\mathcal{G}$-ideals is similar. □

Corollary 2.3.6. *The thick subcategory of $\mathcal{G}$-finite objects is essentially small.* □

We now return to the problem of constructing objects in a stable homotopy category from a given set of small objects.

Definition 2.3.7. Let $\mathcal{A}$ be an essentially small thick subcategory of small objects in $\mathcal{C}$. Let $X \in \mathcal{C}$ be an arbitrary object. Write $\Lambda_{\mathcal{A}}(X)$ for the category whose objects are maps $(Z \xrightarrow{u} X)$ with $Z \in \mathcal{A}$, and whose morphisms are maps $Z \xrightarrow{v} Z'$ such that $u'v = u$. This again has only a set of isomorphism classes. We often write X_α for a typical object of $\Lambda_{\mathcal{A}}(X)$. If $H\colon \mathcal{A} \longrightarrow \mathrm{Ab}$ is an exact functor, write

$$\widehat{H}_{\mathcal{A}}(X) = \varinjlim_{\Lambda_{\mathcal{A}}(X)} H(X_\alpha).$$

Note that $\widehat{H}_{\mathcal{A}}$ is a functor on all of $\mathcal{C}$. If $\mathcal{C}$ is algebraic, we can let $\mathcal{A}$ be the category of $\mathcal{G}$-finite objects (which is essentially small by Corollary 2.3.6). In this case, we write $\Lambda(X)$ for $\Lambda_{\mathcal{A}}(X)$ and $\widehat{H}(X)$ for $\widehat{H}_{\mathcal{A}}(X)$.

Before stating the properties of $\Lambda_{\mathcal{A}}(X)$ and $\widehat{H}_{\mathcal{A}}$, we need to recall a definition.

Definition 2.3.8. A category $\mathcal{I}$ is *filtered* if

1. For any $i, j \in \mathcal{I}$ there exists an object $k \in \mathcal{I}$ and maps $i \longrightarrow k \longleftarrow j$.
2. Given any two maps $u, v\colon i \longrightarrow j$ in $\mathcal{I}$, there is an object $k \in \mathcal{I}$ and a map $w\colon j \longrightarrow k$ with $wu = wv$.

A functor $F\colon \mathcal{J} \longrightarrow \mathcal{I}$ of filtered categories is *cofinal* if

1. For any $i \in \mathcal{I}$ there exists an object $j \in \mathcal{J}$ and a map $u\colon i \longrightarrow Fj$.
2. Given any two maps $u, v\colon i \longrightarrow Fj$ in $\mathcal{I}$, there is an object $k \in \mathcal{J}$ and a map $w\colon j \longrightarrow k$ with $(Fw)u = (Fw)v$.

If $\mathcal{I}$ is filtered, then it is well-known and easy to see that the colimit functor from $\mathcal{I}$-indexed diagrams of Abelian groups to Abelian groups is exact. If $F\colon \mathcal{J} \longrightarrow \mathcal{I}$ is cofinal, and $A\colon \mathcal{I} \longrightarrow \mathrm{Ab}$, then it is also well-known that

$$\varinjlim_{\mathcal{I}} A_i = \varinjlim_{\mathcal{J}} A_{Fj}.$$

Proposition 2.3.9. *$\Lambda_{\mathcal{A}}(X)$ is a filtered category, functorial in X. It has a terminal object if X lies in $\mathcal{A}$. Moreover, $\widehat{H}_{\mathcal{A}}$ is a homology functor, which agrees with H on $\mathcal{A}$.*

Proof. $\mathcal{A}$ is triangulated, so it has finite weak colimits. These can also be used as finite weak colimits in $\Lambda_{\mathcal{A}}(X)$, so $\Lambda_{\mathcal{A}}(X)$ is a filtered category.

Suppose we have a map $f\colon X \longrightarrow Y$. This gives an evident functor $\Lambda_{\mathcal{A}}(X) \longrightarrow \Lambda_{\mathcal{A}}(Y)$, sending $(U \xrightarrow{u} X)$ to $(U \xrightarrow{fu} Y)$. If $X \in \mathcal{A}$ then it is immediate that $(X \xrightarrow{1} X)$ is the terminal object in $\Lambda_{\mathcal{A}}(X)$.

Next we show that $\widehat{H}_{\mathcal{A}}$ is additive. Suppose that $U \in \mathcal{A}$ and $a \in H(U)$. Then, for any map $u\colon U \longrightarrow X$ we get an object $(U,u) = (U \xrightarrow{u} X)$ of $\Lambda_{\mathcal{A}}(X)$, and thus an element of $\widehat{H}_{\mathcal{A}}(X)$, which we shall call $[u,a]$. Suppose that we have two different maps $u,v\colon U \longrightarrow X$; we claim that $[u+v,a] = [u,a] + [v,a]$. To see this, consider the following diagram in $\Lambda_{\mathcal{A}}(X)$.

$$\begin{array}{ccccc} (U,u) & \xrightarrow{i_0} & (U \amalg U, u \amalg v) & \xleftarrow{i_1} & (U,v) \\ & & \Delta \big\uparrow & & \\ & & (U, u+v) & & \end{array}$$

It is clear that

$$[u+v,a] = [u \amalg v, \Delta_* a] = [u \amalg v, (a,a)] = [u \amalg v, (a,0)] + [u \amalg v, (0,a)].$$

Similarly, we have $[u,a] = [u \amalg v, i_{0*}a] = [u \amalg v, (a,0)]$ and $[U,v] = [u \amalg v, (0,a)]$, which proves the claim.

It follows that $\widehat{H}_{\mathcal{A}}$ is an additive functor. Thus, when I is finite we have

$$\widehat{H}_{\mathcal{A}}\left(\coprod_{i\in I} X_i\right) = \bigoplus_{i\in I} \widehat{H}_{\mathcal{A}}(X_i) \tag{2.3.1}$$

On the other hand, if I is infinite then one sees (using the smallness of objects of $\mathcal{A}$) that

$$\widehat{H}_{\mathcal{A}}\left(\coprod_{i\in I} X_i\right) = \varinjlim{}_J \widehat{H}_{\mathcal{A}}\left(\coprod_{i\in J} X_i\right)$$

where J runs over finite subsets of I. It follows that (2.3.1) holds even when I is infinite; i.e., $\widehat{H}_{\mathcal{A}}$ takes arbitrary coproducts to direct sums.

We now show that $\widehat{H}_{\mathcal{A}}$ is an exact functor. Let $X \xrightarrow{f} Y \xrightarrow{g} Z$ be a cofiber sequence. Suppose that $y' \in \widehat{H}_{\mathcal{A}}(Y)$. Then there is an object $V \xrightarrow{v} Y$ of $\Lambda_{\mathcal{A}}(Y)$ and an element $y \in H(V)$ which represents y. Now suppose that y' maps to zero in $\widehat{H}_{\mathcal{A}}(Z)$. By examining the definitions, we see that there is a factorization $(V \xrightarrow{v} Y \xrightarrow{f} Z) = (V \xrightarrow{m} W \xrightarrow{w} Z)$ such that $W \in \mathcal{A}$ and $H(m)(y) = 0$. Let $U \xrightarrow{k} V$ be the fiber of m, so we can choose a map $u\colon U \longrightarrow X$ making the following diagram commute:

$$\begin{array}{ccccc} U & \xrightarrow{k} & V & \xrightarrow{m} & W \\ u\big\downarrow & & v\big\downarrow & & w\big\downarrow \\ X & \xrightarrow{f} & Y & \xrightarrow{g} & Z \end{array}$$

Because $H(m)(y) = 0$, we see that $y = H(k)(x)$ for some $x \in H(U)$. This defines an element $x' \in \widehat{H}_{\mathcal{A}}(X)$, whose image in $\widehat{H}_{\mathcal{A}}(Y)$ is y'. This means that $\widehat{H}_{\mathcal{A}}$ is an exact functor, and in fact a homology functor.

If we restrict to $\mathcal{A}$, there is an evident natural transformation $\widehat{H}_{\mathcal{A}}(X) \longrightarrow H(X)$. For $X \in \mathcal{A}$, the colimit of H over $\Lambda_{\mathcal{A}}(X)$ is just the value at the terminal object, in other words $H(X)$. □

Remark 2.3.10. If H and $\mathcal{A}$ are as in Definition 2.3.7, and if H' is any homology functor extending H, there is a natural transformation $\widehat{H}_{\mathcal{A}} \longrightarrow H'$ of homology functors that is an isomorphism on the localizing subcategory $\operatorname{loc}\langle\mathcal{A}\rangle$ generated by $\mathcal{A}$.

For convenience, we record the most important special case separately.

Corollary 2.3.11. *Let $\mathcal{C}$ be an algebraic stable homotopy category, and $\mathcal{F}$ the subcategory of $\mathcal{G}$-finite objects. If H is an exact functor $\mathcal{F} \longrightarrow \mathrm{Ab}$, then*

$$\widehat{H}(X) = \varinjlim_{\Lambda(X)} H(X_\alpha)$$

defines a homology functor $\mathcal{C} \longrightarrow \mathrm{Ab}$ extending H. Moreover, any other extension of H is canonically isomorphic to $\widehat{H}$. □

We can finally finish the proof of Theorem 2.1.3

Corollary 2.3.12. *An object X in an algebraic stable homotopy category is small if and only if it is $\mathcal{G}$-finite.*

Proof. Suppose that X is small, so that $[X,-]$ is a homology theory. It follows by Proposition 2.3.9 that $[X,X] = \varinjlim_{\Lambda(X)}[X,X_\alpha]$. In particular, $1_X \in [X,X]$ must factor through some $X_\alpha \in \Lambda(X)$, in other words X is a retract of X_α. By the definition of $\Lambda(X)$, X_α is $\mathcal{G}$-finite, so X is $\mathcal{G}$-finite. We have already seen the converse in Theorem 2.1.3. □

We would like to prove an analogous result for cohomology functors, but the failure of the inverse limit functor to be exact prevents us from doing so in general. However, there is one situation when the inverse limit is exact.

Definition 2.3.13. Given a ring R, a *linear topology* on an R-module M is a topology such that the cosets $U + m$ of open submodules U form a basis of open sets. A module M with a linear topology is *linearly compact* if it is Hausdorff, and for every family of closed cosets $\{A_\alpha\}$, the intersection $\bigcap A_\alpha$ is empty if and only if some finite subintersection is empty (see for example [Jen72, p. 56]). In particular, if M is compact Hausdorff then it is linearly compact.

Note that products and closed subspaces of linearly compact modules are linearly compact [Jen72], so the inverse limit over a filtered category of linearly compact modules under continuous maps is again linearly compact. Moreover, [Jen72, Théorème 7.1] implies that the inverse limit functor (taken over any filtered category) is exact on the category of linearly compact R-modules and continuous homomorphisms. Jensen only states this theorem for inverse limits over directed sets, but the proof works for filtered categories as well. However, we give a different proof here, because we need its stronger statement in what follows.

Proposition 2.3.14. *Suppose that R is a ring, $\mathcal{I}$ is a filtered category, and $M\colon \alpha \mapsto M_\alpha$ is a functor from $\mathcal{I}$ to the category of linearly compact R-modules and continuous homomorphisms. Suppose that for each object α of $\mathcal{I}$, C_α is a (necessarily nonempty) closed coset of M_α such that C forms a subfunctor of M. (That is, for any map $\alpha \longrightarrow \beta$, the induced map $M_\beta \longrightarrow M_\alpha$ takes C_β to C_α.) Then $\varprojlim C_\alpha$ is nonempty.*

Proof. As mentioned above, $\prod M_\alpha$ is linearly compact. Given any finite collection $J \neq \emptyset$ of morphisms of $\mathcal{J}$, let D denote the set of domains of $r \in J$ and R the set of codomains of $r \in J$. For each $\beta \in D \cup R$, let T_β denote the subset of $\prod M_\alpha$ consisting of all (m_α) such that $m_\beta \in C_\beta$. Then T_β is a closed coset in $\prod M_\alpha$. For each morphism $r\colon \beta \to \gamma \in J$, let U_r denote the subset of $\prod M_\alpha$ consisting of all (m_α) such that $r(m_\gamma) = m_\beta$. Then U_r is the kernel of the continuous homomorphism

$$\prod M_\alpha \longrightarrow M_\beta \times M_\gamma \xrightarrow{(1,-r)} M_\beta.$$

In particular, U_r is a closed subgroup.

Now let $S_J = \bigcap_{\alpha \in D \cup R} T_\alpha \cap \bigcap_{r \in J} U_r$. We claim that $S_J \neq \emptyset$. Because $\mathcal{J}$ is filtered, there is an object γ and maps $\alpha \xrightarrow{s_\alpha} \gamma$ for every $\alpha \in D \cup R$ such that $s_\alpha = s_\beta \circ r$ for every $r\colon \alpha \to \beta$ in J. One can now choose a class $c \in C_\gamma$ and define $m_\alpha = M_{s_\alpha}(c)$ for $\alpha \in D \cup R$ and 0 otherwise. Clearly $(m_\alpha) \in S_J$, so S_J is nonempty as claimed. As it is a nonempty intersection of closed cosets, it is a closed coset. Note that any finite intersection of S_J's is again an S_J. Thus, by linear compactness, the intersection of the S_J's is nonempty. This intersection is precisely $\varprojlim C_\alpha$. □

The exactness of the inverse limit on linearly compact modules is then immediate:

Corollary 2.3.15. *Let R be a ring, and let $\mathcal{J}$ be a filtered category. Consider the (Abelian) category $[\mathcal{J}^{\text{op}}, \mathcal{M}]$ of contravariant functors from $\mathcal{J}$ to the category $\mathcal{M}$ of linearly compact R-modules and continuous homomorphisms. Then the inverse limit functor $[\mathcal{J}^{\text{op}}, \mathcal{M}] \to \mathcal{M}$ is exact.*

Proof. Suppose that $M_\alpha \xrightarrow{f_\alpha} N_\alpha \xrightarrow{g_\alpha} P_\alpha$ is an exact sequence of inverse systems of linearly compact R-modules. Suppose $(n_\alpha) \in \varprojlim N_\alpha$, and $g_\alpha n_\alpha = 0$ for all α. Let $C_\alpha = f_\alpha^{-1}\{n_\alpha\}$. Then C_α is nonempty, and thus a closed coset in M_α. By the preceding proposition, $\varprojlim C_\alpha$ is nonempty. Any element in it is a class in $\varprojlim M_\alpha$ mapping to (n_α). □

We can now prove the following proposition.

Proposition 2.3.16. *Suppose that $\mathcal{A}$ is an essentially small thick subcategory of small objects in a stable homotopy category $\mathcal{C}$. Let R be a ring, and let $\mathcal{M}$ be the category of linearly compact R-modules and continuous homomorphisms. Suppose that $H\colon \mathcal{A}^{\text{op}} \to \mathcal{M}$ is an exact functor. Then*

$$\widehat{H}_{\mathcal{A}}(X) = \varprojlim_{\Lambda_{\mathcal{A}}(X)} H(X_\alpha)$$

defines a cohomology functor $\mathcal{C}^{\text{op}} \to \mathcal{M}$, which agrees with H on $\mathcal{A}$.

Proof. We first claim that $\widehat{H}$ takes coproducts to products. The proof does not use linear compactness, and is similar to the proof of the analogous part of Proposition 2.3.9 (see also [Mar83, Proposition 4.8]).

It therefore suffices to check that $\widehat{H}$ is exact. Let $X \xrightarrow{f} Y \xrightarrow{g} Z$ be a cofiber sequence. Define $\Lambda_{\mathcal{A}}(g)$ to be the category of commutative squares

$$\begin{array}{ccc} U & \longrightarrow & V \\ \downarrow & & \downarrow \\ Y & \xrightarrow{g} & Z \end{array}$$

where U and V are in a small skeleton of $\mathcal{A}$. The morphisms are commutative diagrams. There are then obvious functors from $\Lambda_{\mathcal{A}}(g)$ to $\Lambda_{\mathcal{A}}(Y)$ and $\Lambda_{\mathcal{A}}(Z)$, and it is straightforward to verify that these are cofinal. We write $Y_\alpha \longrightarrow Z_\alpha$ for a typical object of $\Lambda_{\mathcal{A}}(g)$. Thus $\widehat{H}(Y) = \varprojlim_{\Lambda_{\mathcal{A}}(g)} H(Y_\alpha)$, and similarly for $\widehat{H}(Z)$.

Now suppose that we are given a class $y \in \widehat{H}(Y)$ such that $\widehat{H}(f)y = 0$. The class y is given by a compatible family $y_\alpha \in H(Y_\alpha)$ for each $\alpha \in \Lambda_{\mathcal{A}}(g)$. Then for each $\alpha \in \Lambda_{\mathcal{A}}(g)$, we have a map $Y_\alpha \xrightarrow{g_\alpha} Z_\alpha$, so we can let $C_\alpha = H(g_\alpha)^{-1}(y_\alpha)$. We claim that C_α is nonempty. Indeed, let $X_\alpha \xrightarrow{f_\alpha} Y_\alpha$ denote the fiber of g_α. Then $X_\alpha \in \mathcal{A}$, and the induced map $X_\alpha \longrightarrow Y$ factors through X. Therefore, since $\widehat{H}(f)y = 0$, we must have $H(f_\alpha)(y_\alpha) = 0$, so, by the exactness of H, C_α is nonempty. It is then clear that C_α is a closed coset, so, by Proposition 2.3.14, $\varprojlim C_\alpha$ is nonempty. A class in this inverse limit is a $z \in \widehat{H}(Z)$ such that $\widehat{H}(g)(z) = y$. □

We now discuss when an object X can be constructed from a given set of finite objects.

Proposition 2.3.17. *Suppose that $\mathcal{B} = \{F_i\}$ is a set of small objects in a stable homotopy category $\mathcal{C}$. Let $\mathcal{A} = \operatorname{thick}\langle\mathcal{B}\rangle$ be the thick subcategory generated by $\mathcal{B}$, which is essentially small by Proposition 2.3.5. Consider the following conditions on an object X of $\mathcal{C}$:*

(a) *X is the sequential colimit of a sequence*

$$0 = X^0 \longrightarrow X^1 \longrightarrow X^2 \longrightarrow \dots$$

such that the cofiber of $X^k \longrightarrow X^{k+1}$ is a coproduct of suspensions of elements of $\mathcal{B}$.

(b) *X is in the localizing subcategory $\operatorname{loc}\langle\mathcal{B}\rangle$ generated by $\mathcal{B}$.*

(c) *For every homology functor H defined on $\mathcal{C}$, the natural map $\widehat{H}_{\mathcal{A}}(X) \longrightarrow H(X)$ is an isomorphism.*

Then we have (a)⇔(b)⇒(c), and if $\mathcal{C}$ is algebraic then (c)⇒(b). Moreover, a small object X is in $\operatorname{loc}\langle\mathcal{B}\rangle$ if and only if it lies in $\mathcal{A}$.

Proof. Write $\mathcal{D} = \operatorname{loc}\langle\mathcal{A}\rangle = \operatorname{loc}\langle\mathcal{B}\rangle$.

We first make a construction for arbitrary $X \in \mathcal{C}$. This is very similar to the construction in Proposition 2.3.1, to which we refer for more details. Let $X_0 = X$ and let A_0 be the coproduct

$$\coprod_{Z \in \Sigma^*\mathcal{B}} \coprod_{f \in [Z, X_0]} Z.$$

There is an obvious map $A_0 \longrightarrow X_0$ which induces a surjection $[Z, A_0] \longrightarrow [Z, X_0]$ for all $Z \in \mathcal{A}$. Let X_1 be the cofiber of this map. Iterating, we get cofibrations $A_k \longrightarrow X_k \longrightarrow X_{k+1}$ in which A_k is a coproduct of suspensions of copies of objects in $\mathcal{B}$, and the map $[F_i, X_j]_* \longrightarrow [F_i, X_{j+1}]_*$ is zero for every F_i.

Now let X^k be the fiber of the map $X \longrightarrow X_k$, and CX the sequential colimit of the X^k, so we get a map $CX \longrightarrow X$. As in Proposition 2.3.1, we see that $[Z, CX] \simeq [Z, X]$ for all $Z \in \mathcal{A}$. Let LX be the cofiber of $CX \longrightarrow X$, so that $[Z, LX] = 0$ for all $Z \in \mathcal{A}$. As the category of those Z for which $[Z, LX]_* = 0$ is localizing, we conclude that $[Z, LX] = 0$ for all $Z \in \mathcal{D}$.

We now turn to the main part of the proof. It is clear that (a)⇒(b).

Suppose that (b) holds. We claim that $LX = 0$. Indeed, $LX \in \mathcal{D}$ and $[Z, LX] = 0$ when $Z \in \mathcal{D}$, so $[LX, LX] = 0$, so $LX = 0$. Thus $X = CX$, and so (a) holds.

Suppose again that (a) holds, and that $H\colon \mathcal{C} \to \mathrm{Ab}$ is a homology functor. To prove that (a)$\Rightarrow$(c), we need to show that $\widehat{H}_{\mathcal{A}}(X) = H(X)$. Both $\widehat{H}_{\mathcal{A}}$ and H are homology functors, so the subcategory $\mathcal{D}'$ on which the map $\widehat{H}_{\mathcal{A}} \to H$ is an isomorphism, is localizing. As $\mathcal{A} \subseteq \mathcal{D}'$, we see that $\mathcal{D} \subseteq \mathcal{D}'$, in particular $X \in \mathcal{D}'$. Thus (a)$\Rightarrow$(c).

Now suppose that $\mathcal{C}$ is algebraic, and that (c) holds. Let H be a homology theory. Because $[Z, LX] = 0$ when $Z \in \mathcal{A}$, we see that $\widehat{H}_{\mathcal{A}}(LX) = 0$, and thus $\widehat{H}_{\mathcal{A}}(CX) = \widehat{H}_{\mathcal{A}}(X)$. On the other hand, $\widehat{H}_{\mathcal{A}}(X) = H(X)$ by assumption, and $\widehat{H}_{\mathcal{A}}(CX) = H(CX)$ because (b)$\Rightarrow$(c). Thus $H(CX) = H(X)$ for every homology theory H. Because $\mathcal{C}$ is algebraic, the functor $[Z, -]$ is a homology theory whenever $Z \in \mathcal{G}$, so $[Z, CX] = [Z, X]$. It follows that $CX = X$, and so (c)$\Rightarrow$(a).

Finally, suppose that X is small and in $\mathcal{D}$. Then $[X, -]$ is a homology theory, so, since (b)$\Rightarrow$(c), $[X, X] = \varinjlim \Lambda(X)[X, X_\alpha]$. It follows that $1 \in [X, X]$ factors through some $X_\alpha \in \mathcal{A}$; in other words X is a retract of X_α, so $X \in \mathcal{A}$. □

Remark 2.3.18.

(a) The map $X \to LX$ constructed in this proof is finite localization away from $\mathcal{A}$. See Theorem 3.3.3 for details.

(b) Suppose that $\mathcal{C}$ is an algebraic stable homotopy category, $\mathcal{A}$ is an essentially small thick subcategory, and $X \in \mathrm{loc}\langle\mathcal{A}\rangle$. Then the natural map $\widehat{H}_{\mathcal{A}}(X) \to H(X)$ is an isomorphism for all homology functors H defined on $\mathcal{A}$, making it look as though X should be the minimal weak colimit of the diagram $\Lambda_{\mathcal{A}}(X)$. Indeed, if $\Lambda_{\mathcal{A}}(X)$ has a minimal weak colimit X', then we have a map $X' \to X$ which gives an isomorphism under any homology functor, so that $[Z, X'] \simeq [Z, X]$ for every $Z \in \Sigma^*\mathcal{G}$. It follows that $X' \simeq X$. In particular, this holds if $\mathcal{C}$ is a Brown category (see Definition 4.1.4 and Theorem 4.2.3).

3. Bousfield localization

We now present some basic results on Bousfield localization in a stable homotopy category; in particular, we define localization functors and investigate their properties, and we show that, in an algebraic stable homotopy category, one can localize with respect to any homology functor. We also discuss the properties of the full subcategory of L-local objects, for various kinds of localization functors L.

The first definition of this kind of localization was given by Adams [Ada74]. Certain set-theoretic problems with Adams' definition were cured by Bousfield [Bou79a, Bou79b, Bou83]. See [Rav84, Rav92] for an analysis of some particularly important and interesting examples in the homotopy category of spectra.

3.1. Localization and colocalization functors.

See Section A.1 for the definition of an exact functor between triangulated categories.

Definition 3.1.1.

(a) Suppose that $i\colon 1 \to L$ is a natural transformation of exact functors from $\mathcal{C}$ to itself. We say that the pair (L, i), or just L, is a *localization functor* if
  (i) The natural transformation Li from L to L^2 is an equivalence.

(ii) For all objects X, Y the map

$$[LX, LY] \xrightarrow{i_X^*} [X, LY]$$

is an isomorphism.

(iii) If $LX = 0$ then $L(X \wedge Y) = 0$ for all Y.

(b) Dually, suppose that $q\colon C \longrightarrow 1$ is a natural transformation of exact functors from $\mathcal{C}$ to itself. We say that (C, q), or just C, is a *colocalization functor* if

(i) The natural transformation Cq from C^2 to C is an equivalence.

(ii) For all objects X, Y the map

$$[CX, CY] \xrightarrow{q_{Y*}} [CX, Y]$$

is an isomorphism.

(iii) If $CX = 0$ then $CF(Y, X) = 0$ for all Y.

(c) A *morphism of localization functors* is a natural transformation $u\colon L \longrightarrow L'$ of exact functors such that $ui = i'$ (and similarly for colocalization functors).

(d) If $M\colon \mathcal{C} \longrightarrow \mathcal{C}$ is an exact functor and MX is trivial, we say that X is *M-acyclic*. A map $X \xrightarrow{f} Y$ is called an *M-equivalence* if Mf is an isomorphism.

(e) If L is a localization functor and $i_X : X \longrightarrow LX$ is an isomorphism, we say that X is *L-local*; we let $\mathcal{C}_L$ denote the full subcategory of L-local objects.

(f) Dually, if C is a colocalization functor and $j_X : CX \longrightarrow X$ is an isomorphism, we say that X is *C-colocal*; we let $\mathcal{C}^C$ denote the full subcategory of C-colocal objects.

We start by proving a basic fact.

Lemma 3.1.2. *Let L be a localization functor on a stable homotopy category $\mathcal{C}$. Then an object $Y \in \mathcal{C}$ is L-local if and only if $Y \simeq LX$ for some X. A similar statement holds for colocalization functors.*

Proof. If Y is local, then $Y \simeq LY$. For the converse, we need only prove that LX is local for all X. By naturality of i, we have

$$(Li_X) \circ i_X = i_{LX} \circ i_X : X \longrightarrow L^2X.$$

It follows from condition (ii) that $Li_X = i_{LX}$, so i_L is an equivalence as required. □

Remark 3.1.3. Much of the theory can be developed without conditions (a)(iii) and (b)(iii), but we do not know of interesting applications for this. We have therefore included these conditions so as to simplify the hypotheses for various results below.

Remark 3.1.4. We have followed the usual practice in category theory by defining colocalization functors to be the dual thing to localization functors. Unfortunately, this conflicts with the language used by Bousfield in [Bou79b, Bou79a]. His "colocalization with respect to E" is actually a localization functor in our language. The associated category of acyclics is the localizing subcategory $\mathrm{loc}\langle E\rangle$ generated by E.

It turns out that there are rather few morphisms of localization functors.

Lemma 3.1.5. *Let L and L' be localization functors on a stable homotopy category $\mathcal{C}$. If $iL'\colon L' \longrightarrow LL'$ is an isomorphism (equivalently, if L'-local objects are L-local), then there is a unique morphism of localization functors $u = (iL')^{-1} \circ Li'$ from L to L'; otherwise, there are no such morphisms.*

Proof. Firstly, suppose that iL' is an isomorphism. (By Lemma 3.1.2, it is equivalent to say that $\mathcal{C}_{L'} \subseteq \mathcal{C}_L$.) It is easy to check that $u = (iL')^{-1} \circ Li'$ is a morphism from L to L'. Suppose that v is another such morphism. Then $ui = i' = vi$ but $L'X$ is L-local, so $i^* \colon [LX, L'X] \to [X, L'X]$ is an isomorphism, so $u = v$.

Conversely, suppose that we have a morphism $u\colon L \to L'$. First, we claim that

$$(iL') \circ u = Li' \colon L \to LL'.$$

As the target is L-local, it is enough to check this after composing with i. Using the naturality of i, we find that

$$(iL') \circ u \circ i = (iL') \circ i' = (Li') \circ i,$$

as required.

Next, consider the composite

$$w = (LL' \xrightarrow{uL'} L'L' \xrightarrow{(i'L')^{-1}} L').$$

It is easy to see that the composite $L'X \xrightarrow{iL'} LL'X \xrightarrow{w} L'X$ is the identity, so $L'X$ is a retract of the L-local object $LL'X$ and thus is L-local. Thus L'-local objects are L-local. □

Lemma 3.1.6. *Suppose that $\mathcal{C}$ is a stable homotopy category.*

(a) *There is a natural equivalence between localization and colocalization functors, in which L and C correspond if and only if $CX \to X \to LX$ is a cofiber sequence. More precisely, consider the following category $\mathcal{B}$. An object consists of exact functors (C, L) and morphisms (q, i, d) of exact functors such that (C, q) is a colocalization, (L, i) is a localization, and*

$$CX \xrightarrow{q} X \xrightarrow{i} LX \xrightarrow{d} \Sigma CX$$

is exact. The morphisms are the evident thing. Then the forgetful functors from $\mathcal{B}$ to the categories of localization and colocalization functors are equivalences.

In the following statements, we assume that C and L correspond as above.

(b) *The following are equivalent:*
  (i) *X is L-local.*
  (ii) *$i_X \colon X \to LX$ is an equivalence.*
  (iii) *$X \simeq LY$ for some Y.*
  (iv) *$[Z, X] = 0$ (or $[Z, X]_* = 0$, or $F(Z, X) = 0$) for all L-acyclic (equivalently, C-colocal) objects Z.*
  (v) *X is C-acyclic.*

(Of course, (i) ⇔ (ii) by definition.) Dually, the following are equivalent:
  (i) *X is C-colocal.*
  (ii) *$q_X \colon CX \to X$ is an equivalence.*
  (iii) *$X \simeq CY$ for some Y.*
  (iv) *$[X, Z] = 0$ (or $[X, Z]_* = 0$, or $F(X, Z) = 0$) for all C-acyclic (equivalently, L-local) objects Z.*
  (v) *X is L-acyclic.*

(c) *$i_X \colon X \to LX$ is initial among L-local objects under X, and terminal among L-equivalences out of X. Dually, $q_X \colon CX \to X$ is terminal among L-acyclic objects over X, and initial among C-equivalences into X.*

(d) *The class of L-acyclics ($= C$-colocals) forms a localizing ideal, and the class of L-locals ($=C$-acyclics) forms a colocalizing coideal.*

(e) *As a functor from $\mathcal{C}$ to $\mathcal{C}_L$, L is left adjoint to the inclusion of the L-local objects into $\mathcal{C}$; similarly, C is right adjoint to the inclusion of the C-colocal objects into $\mathcal{C}$. In particular, L and C are uniquely determined by the subcategory of L-acyclics, or by the subcategory of L-locals.*

Proof. First we observe that if X is L-acyclic and Y is L-local, then $[X,Y]_* = 0$. Indeed, $LY = Y$ and $LX = 0$ and $i^* \colon [LX, LY]_* \simeq [X, LY]_*$ so

$$[X,Y]_* = [X,LY]_* = [LX,LY]_* = 0.$$

Also, we recall from Lemma 3.1.2 that LX is always L-local.

(a): Suppose that (L, i) is a localization functor. For each X we can choose a cofiber sequence

$$CX \xrightarrow{q} X \xrightarrow{i} LX \xrightarrow{d} \Sigma CX.$$

By applying L, we see that $LCX = 0$. On the other hand, suppose that $LU = 0$. Then $[U, LX]_* = 0$, so by applying $[U, -]_*$ to the above sequence we find that $[U, CX] \simeq [U, X]$. It follows that CX is terminal among L-acyclic objects over X. Next, consider a morphism $f \colon X \longrightarrow Y$. The axioms for a triangulated category give a commutative diagram as follows, in which g is *a priori* not uniquely determined.

$$\begin{array}{ccccccc}
\Sigma^{-1}LX & \xrightarrow{d} & CX & \xrightarrow{q} & X & \xrightarrow{i} & LX \\
{\scriptstyle Lf}\downarrow & & {\scriptstyle g}\downarrow & & \downarrow{\scriptstyle f} & & \downarrow{\scriptstyle Lf} \\
\Sigma^{-1}LY & \xrightarrow{d} & CY & \xrightarrow{q} & Y & \xrightarrow{i} & LY
\end{array}$$

However, because of the universal property of $CY \xrightarrow{q} Y$, we see that g is unique after all. As it is unique, it is clearly functorial, and we can call it Cf. (If we simply used the universal property to produce g in the first place, we would not know that the left square commutes.)

As L is an exact functor, it comes equipped with an equivalence $L\Sigma \simeq \Sigma L$. As i is a morphism of exact functors, we have $i\Sigma = \Sigma i$ under this identification. It is standard that $LX \xrightarrow{-d} \Sigma CX \xrightarrow{\Sigma q} \Sigma X \xrightarrow{\Sigma i} \Sigma LX$ is a cofiber sequence. An argument similar to the above shows that there is a unique natural equivalence $C\Sigma \simeq \Sigma C$ making the following diagram commute.

$$\begin{array}{ccccccc}
\Sigma^{-1}L\Sigma X & \xrightarrow{\Sigma^{-1}d\Sigma} & C\Sigma X & \xrightarrow{q\Sigma} & \Sigma X & \xrightarrow{\Sigma i} & L\Sigma X \\
\downarrow & & \downarrow & & \downarrow{\scriptstyle 1} & & \downarrow \\
LX & \xrightarrow{-d} & \Sigma CX & \xrightarrow{\Sigma q} & \Sigma X & \xrightarrow{\Sigma i} & \Sigma LX
\end{array}$$

We now show that C is exact. Suppose that

$$X \xrightarrow{f} Y \xrightarrow{g} Z \xrightarrow{h} \Sigma X$$

is a cofiber sequence, and let Z' denote the cofiber of $Cf\colon CX \longrightarrow CY$. Since L is exact, Z' is L-acyclic. We have a morphism of exact sequences

$$\begin{array}{ccccccc} CX & \xrightarrow{Cf} & CY & \xrightarrow{\alpha} & Z' & \xrightarrow{\beta} & \Sigma CX \\ {\scriptstyle q}\downarrow & & {\scriptstyle q}\downarrow & & {\scriptstyle r}\downarrow & & {\scriptstyle \Sigma q}\downarrow \\ X & \xrightarrow{f} & Y & \xrightarrow{g} & Z & \xrightarrow{h} & \Sigma X \end{array}$$

where we have identified ΣCX with $C(\Sigma X)$ as above.

Applying the 5-lemma, we see that $r_*\colon [W, Z'] \longrightarrow [W, Z]$ is an isomorphism for all L-acyclic W. Thus Z' is terminal among L-acyclic objects over Z, and we can therefore identify Z' with CZ and r with q. The proof of the functoriality of C then shows that α and β are uniquely determined, and so must be Cg and Ch respectively.

This makes C into an exact functor, and q and d into morphisms of exact functors.

We need to show that C is a colocalization functor. We saw above that $LC = 0$. It then follows from the cofibration $CCX \xrightarrow{qC} CX \xrightarrow{i} LCX = 0$ that qC is an isomorphism, verifying condition (i). We also saw above that if $LCX = 0$, then $[CX, CY] = [CX, Y]$, verifying condition (ii). Finally, suppose that $CX = 0$ and Y is arbitrary; we need to show that $CF(Y, X) = 0$. As $CF(Y, X)$ is terminal among L-acyclics over $F(Y, X)$, it is enough to show that $CF(Y, X) \xrightarrow{q} F(Y, X)$ is the zero map, or equivalently that the adjoint map $Y \wedge CF(X, Y) \longrightarrow X$ is zero. However, $CF(X, Y)$ and hence $Y \wedge CF(X, Y)$ is L-acyclic, and X is L-local, so $[Y \wedge CF(X, Y), X] = 0$ as required.

Next, consider a morphism $u\colon L \longrightarrow L'$. An evident comparison of cofibrations gives a map $v\colon CX \longrightarrow C'X$ compatible with u. The indeterminacy is measured by $[CX, \Sigma^{-1}L'X]$, but Lemma 3.1.5 tells us that $\Sigma^{-1}L'X$ is L-local, so this group is zero. Thus v is unique, and therefore functorial. Similarly, it is compatible with $\Sigma C \simeq C\Sigma$ and $\Sigma C' \simeq C'\Sigma$.

We have now seen that the functor from $\mathcal{B}$ to the category of localization functors is full, faithful and essentially surjective, hence an equivalence. The argument for colocalization functors is dual.

(b): We prove only the first statement; the proof for the second is dual. By definition, (i) is equivalent to (ii), which is equivalent to (iii) by Lemma 3.1.2. If (iii) holds and $LZ = 0$ then $[Z, X] = [Z, LY] = [LZ, LY] = 0$. We also have $L(U \wedge Z) = 0$ for any U, so $[U, F(Z, X)] = [U \wedge Z, X] = 0$. By taking $U = F(Z, X)$, we see that $F(Z, X) = 0$, and therefore $[Z, X]_* = \pi_* F(Z, X) = 0$. Thus (iii) implies (iv), except that we have not yet verified that L-acyclic means the same as C-colocal. If (iv) holds then $0 = q\colon CX \longrightarrow X$ but $Cq\colon CCX \longrightarrow CX$ is an isomorphism, so $CX = 0$; thus (iv) implies (v). Suppose that (v) holds, so $CX = 0$. Then the cofiber sequence $CX \xrightarrow{q} X \xrightarrow{i} LX$ shows that i_X is an isomorphism, so X is L-local. Thus (v) implies (i). By part of the dual, we see that Z is C-colocal if and only if it is L-acyclic, giving the equivalence of the two versions of (iv).

(c): If $f\colon X \rightarrow Y$ and Y is L-local, then $Y = LY$ so $i_X^*\colon [LX, Y] \simeq [X, Y]$, so f factors uniquely through i_X. Thus $i_X\colon X \longrightarrow LX$ is initial among L-local objects under X. Next, suppose that $g\colon X \longrightarrow Z$ is an L-equivalence. By naturality we have $Lg \circ i_X = i_Z \circ g$, so $h \circ g = i_X$ where $h = (Lg)^{-1} \circ i_Z\colon Z \longrightarrow LX$. If also $h' \circ g = i_X$ then $h - h'\colon Z \longrightarrow LX$ factors through the cofiber of g. As this is

L-acyclic it has no nonzero maps to LX, so $h = h'$. It follows that i_X is terminal among L-equivalences out of X. The claims about C are dual.

(d): All the functors in question are exact, so all the categories in question are thick. Using description (iv) above of the L-local objects, it is immediate that $\mathcal{C}_L$ is closed under arbitrary products, and hence a colocalizing subcategory. Similarly, description (iv) of $\mathcal{C}^C$ shows that it is closed under coproducts and hence localizing.

(e): This follows immediately from (c). □

One quite often needs to consider the fiber of a morphism between localization or colocalization functors. This kind of situation is analyzed in the following proposition.

Proposition 3.1.7. *Consider a morphism $u\colon L \longrightarrow L'$ of localization functors. By part* (a) *of Lemma* 3.1.6, *there is a unique morphism $v\colon C \longrightarrow C'$ of the corresponding colocalization functors such that the diagram*

$$\begin{array}{ccccccc}
CX & \xrightarrow{q} & X & \xrightarrow{i} & LX & \xrightarrow{d} & \Sigma CX \\
{\scriptstyle v}\downarrow & & {\scriptstyle 1}\downarrow & & \downarrow{\scriptstyle u} & & \downarrow{\scriptstyle \Sigma v} \\
C'X & \xrightarrow{q'} & X & \xrightarrow{i'} & L'X & \xrightarrow{d'} & \Sigma C'X
\end{array}$$

commutes. Let MX be the fiber of $u\colon LX \longrightarrow L'X$; then M can be made into a functor in a canonical way, and the various composites of the above functors are given by the following table.

$\circ$	L	L'	C	C'	M
L	L	L'	0	M	M
L'	L'	L'	0	0	0
C	0	0	C	C	0
C'	M	0	C	C'	M
M	M	0	0	M	M

(In particular, $FG = GF$ for any two of these functors.) Moreover, there is an octahedral diagram (as in Definition A.1.1) *of natural maps as follows, so in*

particular MX *is also the cofiber of* $CX \xrightarrow{v} C'X$.

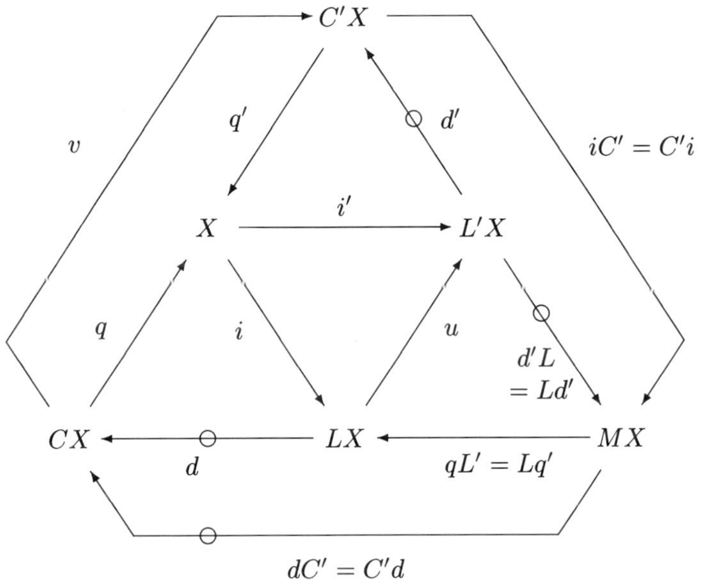

Proof. We saw in Lemma 3.1.5 that $LL' = L'$ and that L'-local objects are L-local. It follows that $CL' = 0$, and thus (using the fibration $C' \longrightarrow 1 \longrightarrow L'$) that $CC' = C$. By a dual argument, $C'C = C$, $L'C = 0$ and $L'L = L'$. After identifying $L' = L'L = LL'$ as above, one finds that $u = Li' = i'L \colon L \longrightarrow L'$ and thus that the fiber MX of u_X is $LC'X = C'LX$. This makes M a functor. More precisely, given a map $f \colon X_0 \longrightarrow X_1$ and cofibrations $M_0 \longrightarrow LX_0 \longrightarrow L'X_0 \longrightarrow \Sigma M_0$ and $M_1 \longrightarrow LX_1 \longrightarrow L'X_1 \longrightarrow \Sigma M_1$, there is a unique map $M_0 \longrightarrow M_1$ compatible with the cofibrations and the maps $Lf, L'f$. Thus, if we choose a fiber MX for each map u_X, then we can make M into a functor in a unique way such that the maps $\Sigma^{-1}L' \longrightarrow M \longrightarrow L$ are natural. The resulting functor is, up to canonical isomorphism, independent of the choices made. One choice is to take $MX = LC'X$ and another is to take $MX = C'LX$, so these are canonically isomorphic. Dually, we have $v = C'q = qC' \colon C \longrightarrow C'$. It follows that the cofiber of v is $C'L = LC'$, which is the same as M again. This justifies all of our table of compositions except for the last row and column. These follow from the rest of the table after substituting $M = C'L = LC'$. The octahedral axiom guarantees the existence of an octahedral diagram of the stated kind, except that the bottommost region might not commute and the arrows marked v and $C'i = iC'$ might be something else. However, one can check by naturality that the bottom region commutes, and that v and $C'i = iC'$ are the *unique* maps with the required commutativity properties. □

We conclude this section by observing that localization functors interact well with the smash product.

Proposition 3.1.8. *For any localization functor L, there is a natural map $LX \wedge LY \longrightarrow L(X \wedge Y)$, which interacts with the isomorphisms $S \wedge X = X = X \wedge S$, $X \wedge Y = Y \wedge X$, and $(X \wedge Y) \wedge Z = X \wedge (Y \wedge Z)$ in the obvious way. In particular, LS is a commutative ring object in $\mathcal{C}$, and every L-local object Y is a module over LS in a natural way.*

Proof. First, observe that the map

$$L(i_X \wedge i_Y)\colon L(X \wedge Y) \longrightarrow L(LX \wedge LY)$$

is an equivalence, as one sees easily using the cofibrations $X \wedge CY \longrightarrow X \wedge Y \longrightarrow X \wedge LY$ and $CX \wedge LY \longrightarrow X \wedge LY \longrightarrow LX \wedge LY$. It follows that the map $LX \wedge LY \longrightarrow L(LX \wedge LY)$ factors uniquely through a map $LX \wedge LY \longrightarrow L(X \wedge Y)$. We leave it to the reader to check that this has the right coherence properties. In particular, we get a multiplication map $LS \wedge LS \longrightarrow L(S \wedge S) = LS$ and a unit map $i_S\colon S \longrightarrow LS$ which make LS into a ring object. Moreover, if Y is L-local, we get a multiplication map

$$LS \wedge Y = LS \wedge LY \longrightarrow L(S \wedge Y) = LY = Y,$$

which makes Y into a module over LS. □

3.2. **Existence of localization functors.** We have so far not touched on one basic question: given a localizing ideal $\mathcal{D}$, when is there a localization functor L such that the category of L-acyclics is precisely $\mathcal{D}$? We know by part (e) of Lemma 3.1.6 that such a functor is essentially unique if it exists.

Definition 3.2.1. If $\mathcal{D}$ is a localizing ideal, and there is a localization functor L such that the category of L-acyclics is precisely $\mathcal{D}$, then we shall write $L_{\mathcal{D}} = L$. If H is a homology or cohomology functor, and $\mathcal{D} = \{X \mid H(X \wedge Y) = 0 \text{ for all } Y\}$, and $L_{\mathcal{D}}$ exists, then we shall write $L_H = L_{\mathcal{D}}$. We will refer to L_H-acyclic and L_H-local objects as *H-acyclic* and *H-local*, respectively. Note that $H(X) = 0$ does not imply that X is H-acyclic if $\mathcal{C}$ is not monogenic.

The first result here is due to Bousfield [Bou79b] (see also [Bou83]), who works in a closed model category. Margolis [Mar83] has given a proof that will work in an arbitrary algebraic stable homotopy category.

See Definitions 1.1.3 and 3.1.1 for the relevant definitions in the following.

Theorem 3.2.2 (Bousfield localization). *For any homology functor $H\colon \mathcal{C} \longrightarrow \mathrm{Ab}$ on an algebraic stable homotopy category $\mathcal{C}$, the localization functor L_H exists.*

Proof. For each $X \in \mathcal{C}$, we need to construct a map $X \longrightarrow LX$ where LX is H-local and the fiber is H-acyclic. The methods of Lemma 3.1.6 will then show that L is automatically a functor and in fact a localization functor. The construction of LX in [Mar83, Chapter 7], applied to the homology functor

$$X \mapsto \bigoplus_{Z \in \mathcal{G}} H(X \wedge Z),$$

relies only on basic properties of triangulated categories and homology functors, together with Corollary 2.3.11 and Brown representability of cohomology functors. □

We do not know of any example of a localizing subcategory for which a localization functor can be proved not to exist. Nick Kuhn has suggested to us that the question of whether localizations always exist may not be decidable using only the usual axioms of set theory. Certainly the proof in [Mar83] involves finding big cardinal numbers, so maybe some of the large cardinal axioms are relevant.

In the homotopy category of spectra, Bousfield has shown [Bou79a] that a localization functor exists for any localizing subcategory generated by a set (not a proper class) of objects. (If the subcategory is generated by $\{E_i\}$ and $E = \coprod_i E_i$ then the localization functor is what Bousfield calls E-colocalization.) The proof probably generalizes to any stable homotopy category derived from a closed model category satisfying suitable axioms.

3.3. **Smashing and finite localizations.** We now define a particularly important special class of localization functors (compare [Rav84, Definition 1.28]). First, we need a lemma.

Lemma 3.3.1. *Suppose that L is a localization functor on a stable homotopy category $\mathcal{C}$. Then there is a natural map $\alpha_X : LS \wedge X \to LX$, which is an isomorphism when X is strongly dualizable.*

Proof. The map is much as in Proposition 3.1.8: by applying L to the cofibration $CS \wedge X \to S \wedge X \to LS \wedge X$, we get an equivalence $LX \simeq L(LS \wedge X)$. We define α_X to be the composite $LS \wedge X \xrightarrow{i} L(LS \wedge X) \simeq LX$. Suppose that X is strongly dualizable. Using part (b)(iv) of Lemma 3.1.6, we see that $LS \wedge X = F(DX, LS)$ is L-local, and thus that α_X is an isomorphism. □

Definition 3.3.2. A localization functor $L : \mathcal{C} \to \mathcal{C}$ is *smashing* if it satisfies the following equivalent conditions:

(a) The natural map $\alpha_X : LS \wedge X \to LX$ defined in Lemma 3.3.1 is an isomorphism for all X.
(b) L preserves coproducts.
(c) The colocalizing subcategory $\mathcal{C}_L$ of L-local objects is also a localizing subcategory.

Proof of equivalence. It is easy to see that (a)⇒(b)⇒(c). Suppose that (c) holds. Given a family of objects $\{X_i\}$, we have a cofiber sequence

$$\coprod CX_i \to \coprod X_i \to \coprod LX_i$$

in which the first term is L-acyclic and the last is L-local. By applying L, we see that $L(\coprod X_i) = \coprod LX_i$, so that (b) holds. Finally, suppose that (b) holds. It is then easy to see that the category of those X for which α_X is an isomorphism, is localizing. By Lemma 3.3.1, it contains $\mathcal{G}$, so it is all of $\mathcal{C}$. □

As a special case of smashing localizations, we have the finite localizations, first considered by Miller [Mil92].

Theorem 3.3.3 (Finite localization). *Suppose that $\{X_i\}$ is a set of small objects in a stable homotopy category $\mathcal{C}$. Let $\mathcal{A}$ denote the $\mathcal{G}$-ideal generated by $\{X_i\}$ and let $\mathcal{D}$ denote the localizing ideal generated by $\{X_i\}$. Then there is a smashing localization functor $L = L^f_{\mathcal{A}}$, which depends only on $\mathcal{A}$, whose acyclics are precisely $\mathcal{D}$. Moreover, the small objects in $\mathcal{D}$ are precisely the objects of $\mathcal{A}$.*

Definition 3.3.4. We refer to localization functors of this type as *finite localizations.* Note that there is a finite localization functor for every essentially small $\mathcal{G}$-ideal of small objects. We refer to $L^f_{\mathcal{A}}$ as *finite localization away from* $\{X_i\}$. If $\mathcal{B}$ is a set of small objects of $\mathcal{C}$ and $\mathcal{A}$ is the $\mathcal{G}$-ideal generated by $\mathcal{B}$, we often abuse notation and write $L^f_{\mathcal{B}}$ for $L^f_{\mathcal{A}}$.

Proof. First, note that, by Theorem 2.1.3(a), $\mathcal{A}$ consists of small objects. In fact, it is easy to see that $\mathcal{A}$ is the thick subcategory generated by the set $\mathcal{B} = \{Z_1 \wedge \dots \wedge Z_r \wedge X_i\}$ where $0 \leq r < \infty$, and $Z_j \in \mathcal{G}$. (In the algebraic case, we can just take $r \leq 1$.) Moreover, $\mathcal{A}$ is essentially small by Proposition 2.3.5. Note as well that $\mathcal{D}$ is the localizing subcategory generated by $\mathcal{B}$, by Lemma 1.4.6.

Applying Proposition 2.3.17 to the set of small objects of $\mathcal{B}$, we construct (for each $X \in \mathcal{C}$) a cofiber sequence $CX \xrightarrow{q} X \xrightarrow{i} LX$, in which CX is in $\mathcal{D}$ and $[Z, LX] = 0$ for $Z \in \mathcal{D}$. It follows that $CX \xrightarrow{q} X$ is terminal among objects of $\mathcal{D}$ over X. It follows in turn that C can be made into a functor, and q into a natural transformation. Moreover q_X is an isomorphism for all $X \in \mathcal{D}$, in particular q_{CX} is an isomorphism. We also see that $CX = 0$ if and only if $[Z, X] = 0$ for all $Z \in \mathcal{D}$. The set of such X is a coideal, because $\mathcal{D}$ is an ideal. By assembling these facts, we conclude that C is a colocalization functor, and that L is the complementary localization functor.

The collection of acyclics for L forms a localizing subcategory containing $\mathcal{B}$, and thus $\mathcal{D}$. On the other hand, if $LX = 0$ then $X = CX$, which lies in $\mathcal{D}$ by construction. Thus $\mathcal{D}$ is precisely the category of L-acyclics.

An object X is L-local if and only if $CX = 0$, if and only if $[Z, X] = 0$ for all $Z \in \mathcal{A}$. Using the fact that these objects Z are small, we see that the category of L-local objects is closed under coproducts, so that L is smashing.

Finally, suppose that X is small and lies in $\mathcal{D}$. Then $X \in \mathcal{A}$ by the last part of Proposition 2.3.17. □

Because the functor $L^f_{\mathcal{A}}$ is smashing, its image is a localizing subcategory. We can thus hope to find another localization functor, whose acyclic category is $L^f_{\mathcal{A}}\mathcal{C}$. We have quite a good understanding of this situation, as expressed by the following theorem. Readers who are familiar with chromatic topology may like to bear the following example in mind.

$$\begin{aligned} F(n) &= \text{a finite type-}n\text{ spectrum} \\ \mathcal{C} &= \text{the } E(n)\text{-local category} \\ \mathcal{A} &= \text{the thick subcategory generated by } L_n F(n) \\ \mathcal{D} &= \text{the } n\text{th monochromatic category} \\ \mathcal{E} &= \text{the } K(n)\text{-local category.} \\ L^f_{\mathcal{A}} &= L_{n-1} \\ L_{\mathcal{A}} &= L_{K(n)}. \end{aligned}$$

Theorem 3.3.5. *Let $\mathcal{A}$ be a $\mathcal{G}$-ideal of small objects in a stable homotopy category $\mathcal{C}$. Suppose that $\mathcal{A}$ is essentially small, consists of strongly dualizable objects, and is closed under the Spanier-Whitehead duality functor D (all of which are automatic* *if $\mathcal{C}$ is algebraic). Write*

$$\mathcal{Z} = \mathcal{A}^{\perp} = \{Y \mid \forall W \in \mathcal{A} \quad [W,Y] = 0\}$$
$$\mathcal{D} = {}^{\perp}\mathcal{Z} = \{X \mid \forall Y \in \mathcal{Z} \quad [X,Y] = 0\}$$
$$\mathcal{E} = \mathcal{Z}^{\perp} = \{X \mid \forall Y \in \mathcal{Z} \quad [Y,X] = 0\}$$

Then there are (co)localization functors

$$C^f_{\mathcal{A}}X \longrightarrow X \longrightarrow L^f_{\mathcal{A}}X$$

$$C_{\mathcal{A}}X \longrightarrow X \longrightarrow L_{\mathcal{A}}X$$

with the following properties.

(a) $L^f_{\mathcal{A}}X = L^f_{\mathcal{A}}S \wedge X$ *and* $C^f_{\mathcal{A}}X = C^f_{\mathcal{A}}S \wedge X$.

(b) $L_{\mathcal{A}}X = F(C^f_{\mathcal{A}}S, X)$ *and* $C_{\mathcal{A}}X = F(L^f_{\mathcal{A}}S, X)$.

(c) $\ker(C^f_{\mathcal{A}}) = \ker(L_{\mathcal{A}}) = \text{image}(L^f_{\mathcal{A}}) = \text{image}(C_{\mathcal{A}}) = \mathcal{Z}$.

(d) $\ker(L^f_{\mathcal{A}}) = \text{image}(C^f_{\mathcal{A}}) = \mathcal{D}$, *and this is the localizing subcategory generated by $\mathcal{A}$.*

(e) $\ker(C_{\mathcal{A}}) = \text{image}(L_{\mathcal{A}}) = \mathcal{E}$, *and this is the colocalizing subcategory generated by*

$$\mathcal{A}' = \{W \wedge U \mid W \in \mathcal{A}\,,\ U \in \mathcal{C}\}.$$

(f) *There are isomorphisms* $L_{\mathcal{A}}C^f_{\mathcal{A}} = L_{\mathcal{A}}$ *and* $C^f_{\mathcal{A}}L_{\mathcal{A}} = C^f_{\mathcal{A}}$.

(g) *The functors* $L_{\mathcal{A}}\colon \mathcal{D} \longrightarrow \mathcal{E}$ *and* $C^f_{\mathcal{A}}\colon \mathcal{E} \longrightarrow \mathcal{D}$ *are mutually inverse equivalences.*

Remark 3.3.6. As with the functor $L^f_{\mathcal{A}}$, if $\mathcal{B}$ is a set of small objects in $\mathcal{C}$ and $\mathcal{A}$ is the $\mathcal{G}$-ideal generated by $\mathcal{B}$, we often abuse notation and write $L_{\mathcal{B}}$ for $L_{\mathcal{A}}$.

Proof. We first justify the comments at the beginning of the statement of the theorem. If $\mathcal{C}$ is algebraic, then Corollary 2.3.6 assures us that $\mathcal{A}$ is essentially small, and Theorem 2.1.3 implies that the objects of $\mathcal{A}$ are strongly dualizable, and that $\mathcal{A}$ is closed under D. We shall first prove (a)–(f), and only verify afterwards that $L_{\mathcal{A}}$ is a localization functor and $C_{\mathcal{A}}$ is a colocalization functor.

Because $\mathcal{A}$ is a $\mathcal{G}$-ideal of dualizable objects closed under D, we see (with an obvious notation) that

$$\mathcal{Z} = \{Y \mid F(\mathcal{A}, Y) = \{0\}\} = \{Y \mid \mathcal{A} \wedge Y = \{0\}\}$$

and thus that $\mathcal{Z}$ is an ideal. It follows that

$$\mathcal{D} = \{X \mid F(X, \mathcal{Z}) = \{0\}\}$$

(which is an ideal), and that

$$\mathcal{E} = \{X \mid F(\mathcal{Z}, X) = \{0\}\}$$

(which is a coideal).

The functors $L^f_{\mathcal{A}}$ and $C^f_{\mathcal{A}}$ were defined in Theorem 3.3.3, where it was also proved that $L^f_{\mathcal{A}}X = L^f_{\mathcal{A}}S \wedge X$; the other half of (a) follows easily, as does the fact that

$$L^f_{\mathcal{A}}S = L^f_{\mathcal{A}}S \wedge L^f_{\mathcal{A}}S$$

and

$$C^f_{\mathcal{A}}S \wedge C^f_{\mathcal{A}}S = C^f_{\mathcal{A}}S.$$

We define $L_{\mathcal{A}}X = F(C^f_{\mathcal{A}}S, X)$ and $C_{\mathcal{A}}X = F(L^f_{\mathcal{A}}S, X)$, so that (b) holds by definition. Clearly these are idempotent exact functors, and the cofibration $C^f_{\mathcal{A}}S \to S \to L^f_{\mathcal{A}}S$ gives rise to natural cofibrations

$$C_{\mathcal{A}}X \to X \to L_{\mathcal{A}}X \to \Sigma C_{\mathcal{A}}X.$$

It follows that $\ker(C_{\mathcal{A}}) = \text{image}(L_{\mathcal{A}})$ and that $\ker(L_{\mathcal{A}}) = \text{image}(C_{\mathcal{A}})$, and similarly that $\ker(C^f_{\mathcal{A}}) = \text{image}(L^f_{\mathcal{A}})$ and that $\ker(L^f_{\mathcal{A}}) = \text{image}(C^f_{\mathcal{A}})$. Thus, we need only prove half of (c),(d) and (e).

It follows from Theorem 3.3.3 that

$$\ker(C^f_{\mathcal{A}}) = \text{image}(L^f_{\mathcal{A}}) = \mathcal{Z},$$

$$\text{image}(C^f_{\mathcal{A}}) = \ker(L^f_{\mathcal{A}}) = \mathcal{D},$$

and that $\mathcal{D}$ is the localizing subcategory generated by $\mathcal{A}$.

(c): All that is left to prove is that $\ker(L_{\mathcal{A}}) = \mathcal{Z}$. Suppose that $Y \in \mathcal{Z}$. Then $\{W \mid F(W,Y) = 0\}$ is a localizing subcategory containing $\mathcal{A}$ and therefore also containing $\mathcal{D} = \text{image}(C^f_{\mathcal{A}})$. In particular, it contains $C^f_{\mathcal{A}}S$, so $L_{\mathcal{A}}Y = 0$. Conversely, suppose that $L_{\mathcal{A}}Y = 0$. For $W \in \mathcal{A}$ we have $C^f_{\mathcal{A}}S \wedge W = C^f_{\mathcal{A}}W = W$, so that

$$[W,Y] = [C^f_{\mathcal{A}}S \wedge W, Y] = [W, F(C^f_{\mathcal{A}}S, Y)] = [W, L_{\mathcal{A}}Y] = 0.$$

Thus $Y \in \mathcal{Z}$.

(d): This was all proved in Theorem 3.3.3, as remarked above.

(e): Suppose that $X \in \mathcal{E}$, so that $F(Y,X) = 0$ for all $Y \in \mathcal{Z} = \text{image}(L^f_{\mathcal{A}})$. In particular, $C_{\mathcal{A}}X = F(L^f_{\mathcal{A}}S, X) = 0$, so that $X = L_{\mathcal{A}}X \in \text{image}(L_{\mathcal{A}})$. Conversely, suppose that $X = L_{\mathcal{A}}X = F(C^f_{\mathcal{A}}S, X)$. Then for $Z \in \mathcal{Z} = \ker(C^f_{\mathcal{A}})$ we have $C^f_{\mathcal{A}}S \wedge Z = 0$ and thus

$$F(Z,X) = F(Z, F(C^f_{\mathcal{A}}S, X)) = F(Z \wedge C^f_{\mathcal{A}}S, X) = 0.$$

It follows that $X \in \mathcal{E}$.

We still need to show that $\mathcal{E}$ is the same as the colocalizing subcategory $\mathcal{E}'$ generated by $\mathcal{A}'$. Consider an object $X = W \wedge U \in \mathcal{A}'$, so that $W \in \mathcal{A}$. We have $C^f_{\mathcal{A}}S \wedge DW = DW$, so

$$L_{\mathcal{A}}(X) = F(C^f_{\mathcal{A}}S, W \wedge U) = F(C^f_{\mathcal{A}}S \wedge DW, U) = F(DW, U) = X.$$

Thus $X \in \mathcal{E}$. It follows that $\mathcal{E}$ is a colocalizing subcategory containing $\mathcal{A}'$, so $\mathcal{E}' \subseteq \mathcal{E}$.

On the other hand, suppose that $X \in \mathcal{E}$, so that $X = F(C^f_{\mathcal{A}}S, X)$. Using an $\mathcal{A}$-based cellular tower for $C^f_{\mathcal{A}}S$, we see that X lies in the colocalizing subcategory generated by $F(\mathcal{A}, X) = D(\mathcal{A}) \wedge X \subseteq \mathcal{A}'$. Thus $X \in \mathcal{E}'$.

(f): We saw in (c) that $\ker(L_{\mathcal{A}}) = \text{image}(L^f_{\mathcal{A}})$. Thus, by applying $L_{\mathcal{A}}$ to the cofibration $C^f_{\mathcal{A}}X \to X \to L^f_{\mathcal{A}}X$, we see that $L_{\mathcal{A}}C^f_{\mathcal{A}} = L_{\mathcal{A}}$. The proof that $C^f_{\mathcal{A}}L_{\mathcal{A}} = C^f_{\mathcal{A}}$ is similar.

(g): On $\mathcal{E} = \text{image}(L_{\mathcal{A}})$ we have $L_{\mathcal{A}}C^f_{\mathcal{A}} = L_{\mathcal{A}} = 1$ (using (f)). Similarly, on $\mathcal{D}$ we have $C^f_{\mathcal{A}}L_{\mathcal{A}} = 1$. The claim follows.

We still need to show that $L_{\mathcal{A}}$ is a localization functor. We have already seen that it is idempotent and exact, so we need only check that $[X, L_{\mathcal{A}}Y] = [L_{\mathcal{A}}X, L_{\mathcal{A}}Y]$, or equivalently that $[C_{\mathcal{A}}X, L_{\mathcal{A}}Y] = 0$. This holds because $C_{\mathcal{A}}X \in \mathcal{Z}$ by (c), and $L_{\mathcal{A}}Y \in \mathcal{E}$ by (e), and $[\mathcal{Z}, \mathcal{E}] = \{0\}$ by the definition of $\mathcal{E}$.

It follows as in Lemma 3.1.6 that $C_{\mathcal{A}}$ is a colocalization functor. □

Theorem 3.3.7 (Algebraic localization)**.** *Let $\mathcal{C}$ be a unital algebraic stable homotopy category, and let T be a set of homogeneous elements in the graded ring π_*S. Then there is a finite localization functor L_T and a natural equivalence $\pi_*(L_TX) = T^{-1}\pi_*(X)$.*

Proof. Each element $t \in T$ is a map $S^d \longrightarrow S$, say; write S/t for the cofiber, and $\mathcal{A} = \{S/t \mid t \in T\}$. Write $L_T = L^f_{\mathcal{A}}$, and C_T for the corresponding colocalization functor. Because inverting T is a coproduct-preserving exact functor on $\pi_*(S)$-modules, we see that $T^{-1}\pi_*(X)$ is a homology functor of X. It vanishes on $\mathcal{A}$, and thus on C_TX for all X. It follows that $T^{-1}\pi_*(X) = T^{-1}\pi_*(L_TX)$. On the other hand, we know that $[S/t, L_TX] = 0$. By considering the cofibration $S^d \xrightarrow{t} S \longrightarrow S/t$, we conclude that multiplication by t is an isomorphism on $\pi_*(L_TX)$. Thus $T^{-1}\pi_*(L_TX) = \pi_*(L_TX)$. □

In this case, the functor $L_{\mathcal{A}}$ should be thought of as a kind of completion at the ideal generated by T. See part (c) of Lemma 6.3.5 for a more precise statement along these lines.

We now discuss the telescope conjecture. This was first stated by Ravenel in [Rav84], as a conjecture about spectra. It was reformulated in many different ways, and finally shown by Ravenel to be false. Nonetheless, it can be shown that analogues are true in many interesting stable homotopy categories. Incidentally, this conjecture is mislabeled as the smashing conjecture in [Nee92a]. See Definitions 1.4.3, 3.1.1, 3.3.2, and Definition 3.3.4 for the relevant terms.

Definition 3.3.8. Suppose that $\mathcal{C}$ is a stable homotopy category. We shall say that the *telescope conjecture* holds in $\mathcal{C}$ if every smashing localization of $\mathcal{C}$ is a finite localization. If so, there is a one-to-one correspondence between essentially small $\mathcal{G}$-ideals of small objects and smashing localizations. The essentially small hypothesis is automatic in case $\mathcal{C}$ is algebraic.

3.4. Geometric morphisms.

Consider a localization functor L on a stable homotopy category $\mathcal{C}$. Recall that $\mathcal{C}_L$ is the category of L-local objects in $\mathcal{C}$, or equivalently the image of L. In the next section, we shall study the properties of $\mathcal{C}_L$, and of L considered as a functor from $\mathcal{C}$ to $\mathcal{C}_L$. The answer will be that L is a "geometric morphism" from $\mathcal{C}$ to $\mathcal{C}_L$. The purpose of the present section is to explain this concept.

Definition 3.4.1. Let $\mathcal{C}$ and $\mathcal{D}$ be enriched triangulated categories. A *geometric morphism* from $\mathcal{C}$ to $\mathcal{D}$ is an exact functor $L\colon \mathcal{C} \longrightarrow \mathcal{D}$ which admits a right adjoint J, together with natural isomorphisms

$$\alpha\colon S_{\mathcal{D}} \simeq LS_{\mathcal{C}}$$
$$\mu\colon LX \wedge_{\mathcal{D}} LY \simeq L(X \wedge_{\mathcal{C}} Y).$$

The maps μ and α are required to commute in the evident sense with the symmetric monoidal structures on $\mathcal{C}$ and $\mathcal{D}$. If L has a right adjoint J and maps α, μ as above which are not necessarily isomorphisms, we say that L is a *lax geometric morphism.* If $\mathcal{C}$ and $\mathcal{D}$ are stable homotopy categories and $L : \mathcal{C} \longrightarrow \mathcal{D}$ is a (lax) geometric morphism, we say that L is a *(lax) stable morphism* if L takes $\mathcal{G}$-finite objects of $\mathcal{C}$ to $\mathcal{G}$-finite objects of $\mathcal{D}$.

The terminology is stolen from topos theory. It is justified by the fact that a map $X \longrightarrow Y$ of schemes gives rise to a geometric morphism $\mathcal{D}(Y) \longrightarrow \mathcal{D}(X)$ of derived categories (and also a geometric morphism of the corresponding topoi of sheaves). In some special cases, a geometric morphism (of topoi or of stable homotopy categories) will also admit a left adjoint. The functors which arise in equivariant stable homotopy theory from a change of group or universe [LMS86, Chapter II] are all either geometric morphisms or adjoints of geometric morphisms.

We point out that it is straightforward to compose (lax) geometric morphisms and (lax) stable morphisms. We can therefore form a (very large) category of stable homotopy categories, where the morphisms are stable morphisms, or lax stable morphisms if we prefer.

We can now make more precise our claim that the choice of generators is not very important in an algebraic stable homotopy category. Indeed, suppose $\mathcal{C}$ is an algebraic stable homotopy category with two sets of small generators $\mathcal{G}$ and $\mathcal{G}'$. Then the identity functor is a stable isomorphism between $(\mathcal{C}, \mathcal{G})$ and $(\mathcal{C}, \mathcal{G}')$.

Let $L\colon \mathcal{C} \longrightarrow \mathcal{D}$ be a geometric morphism, with right adjoint J. Let X and Y denote objects of $\mathcal{C}$, and U and V objects of $\mathcal{D}$. By juggling adjoints, one can construct natural maps as follows.

$$\begin{aligned}
\beta\colon S_{\mathcal{C}} &\longrightarrow JS_{\mathcal{D}} \\
\nu\colon JU \wedge_{\mathcal{C}} JV &\longrightarrow J(U \wedge_{\mathcal{D}} V) \\
\mu^{\#}\colon LF_{\mathcal{C}}(X,Y) &\longrightarrow F_{\mathcal{D}}(LX, LY) \\
\nu^{\#}\colon JF_{\mathcal{D}}(U,V) &\longrightarrow F_{\mathcal{C}}(JU, JV) \\
\pi\colon X \wedge_{\mathcal{C}} JU &\longrightarrow J(LX \wedge_{\mathcal{D}} U) \\
\pi^{\#}\colon L(X \wedge_{\mathcal{C}} JU) &\longrightarrow LX \wedge_{\mathcal{D}} U \\
\rho\colon F_{\mathcal{C}}(X, JU) &\simeq JF_{\mathcal{D}}(LX, U) \\
\rho^{\#}\colon LF_{\mathcal{C}}(X, JU) &\longrightarrow F_{\mathcal{D}}(LX, U)
\end{aligned}$$

The isomorphism ρ is a sort of internal version of the adjunction $[X, JU] \simeq [LX, U]$. None of the other maps need be isomorphisms. We refrain from listing any of their commutativity and coherence properties.

If L is merely a lax geometric morphism then we can still construct $\mu^{\#}$, ρ and $\rho^{\#}$, but ρ need not be an isomorphism.

Proposition 3.4.2. *Let $L\colon \mathcal{C} \longrightarrow \mathcal{D}$ be a geometric morphism, with right adjoint J. Then L preserves coproducts, and J is an exact functor which preserves products.*

Proof. It is well-known that left adjoints preserve coproducts and right adjoints preserve products. It is proved in [Mar83, Proposition A2.11] that adjoints of exact functors are exact. □

3.5. **Properties of localized subcategories.** In this section we show that any localization of a stable homotopy category is a stable homotopy category.

Theorem 3.5.1. *Suppose that $\mathcal{C}$ is a stable homotopy category, and that $L\colon \mathcal{C} \longrightarrow \mathcal{C}$ is a localization functor. Then $\mathcal{C}_L$ has a natural structure as a stable homotopy category, such that $L\colon \mathcal{C} \longrightarrow \mathcal{C}_L$ is a stable morphism (the right adjoint is the inclusion*

$J\colon \mathcal{C}_L \to \mathcal{C}$). Considered as a functor from $\mathcal{C}$ to $\mathcal{C}_L$, L preserves the following structure:

(a) *cofibrations*
(b) *the smash product and its unit*
(c) *coproducts*
(d) *(minimal) weak colimits, and in particular sequential colimits*
(e) *strong dualizability.*

(Of course, (a) *and* (b) *are part of the claim that L is a geometric morphism.) The inclusion functor $J\colon \mathcal{C}_L \to \mathcal{C}$ preserves the following structure:*

(a) *cofibrations*
(b) *function objects*
(c) *products*
(d) *sequential limits.*

The following maps are isomorphisms:

$$\nu^{\#}\colon JF_L(U,V) \to F(JU,JV)$$
$$\pi^{\#}\colon L(X \wedge JU) \to LX \wedge_L U$$
$$\rho\colon F(X,JU) \to JF_L(LX,U)$$

(where the subscript L indicates structure in $\mathcal{C}_L$). Moreover, $LJ \simeq 1$.

Theorem 3.5.2. *Suppose in addition that $\mathcal{C}$ is algebraic. Then L preserves smallness if and only if L is smashing. Suppose that this holds. Then $\mathcal{C}_L$ is also algebraic, and J is a lax geometric morphism. The following maps are isomorphisms:*

$$\nu\colon JU \wedge_{\mathcal{C}} JV \to J(U \wedge_L V)$$
$$\pi\colon X \wedge_{\mathcal{C}} JU \to J(LX \wedge_L U)$$

(Again, the subscript L indicates structure in $\mathcal{C}_L$). If $\mathcal{C}$ is a Brown category, then $\mathcal{C}_L$ is also a Brown category.

Proof of Theorem 3.5.1. First, we can triangulate $\mathcal{C}_L$ by declaring that $X \to Y \to Z \to \Sigma X$ is a triangle in $\mathcal{C}_L$ if and only if it is a triangle in $\mathcal{C}$ (and X, Y and Z lie in $\mathcal{C}_L$). Using the fact that $\mathcal{C}_L$ is a thick subcategory of $\mathcal{C}$, we see that this makes $\mathcal{C}_L$ into a triangulated category.

Next, we define a smash product $X \wedge_L Y$ on $\mathcal{C}_L$ by $X \wedge_L Y = L(X \wedge Y)$. It is easy to check (using Proposition 3.1.8) that $S_L = LS$ is a unit for this product, and that it makes $\mathcal{C}_L$ into a symmetric monoidal category. If X is arbitrary and Y is L-local, we can see that $F(X,Y) = F(LX,Y)$, and that this object is L-local. Using this, we see that $\mathcal{C}_L$ can be made into a closed symmetric monoidal category by defining $F_L(X,Y) = F(X,Y)$. It is again easy to check that this structure is compatible with the triangulation.

Suppose that $\{X_i\}$ is a family of objects of $\mathcal{C}_L$. Write $\coprod X_i$ for their coproduct in $\mathcal{C}$. For $Y \in \mathcal{C}_L$, we have

$$[L\left(\coprod X_i\right), Y] = [\coprod X_i, Y] = \prod [X_i, Y].$$

It follows that $\coprod_L X_i = L(\coprod X_i)$ is the categorical coproduct in $\mathcal{C}_L$ of the objects X_i.

We see directly from these constructions that L (considered as a functor $\mathcal{C} \longrightarrow \mathcal{C}_L$) preserves cofibrations, the smash product and its unit, and coproducts. We know from part (e) of Lemma 3.1.6 that L is left adjoint to J. It follows that L is a geometric morphism.

If H is a cohomology functor on $\mathcal{C}_L$, then $H \circ L$ is a cohomology functor on $\mathcal{C}$ (because L preserves coproducts). There is therefore an object X of $\mathcal{C}$ and a natural equivalence $[Y, X] \longrightarrow H(LY)$. By choosing Y to be L-acyclic, we find that X is L-local, and therefore represents H as a functor on $\mathcal{C}_L$.

Suppose that $Z \in \mathcal{C}$ is strongly dualizable. We claim that LZ is strongly dualizable in $\mathcal{C}_L$; in other words, for every $Y \in \mathcal{C}_L$ we claim that

$$F_L(LZ, Y) = F_L(LZ, S_L) \wedge_L Y.$$

Indeed, the left hand side is just $F(LZ, Y) = F(Z, Y)$. The right hand side is $L(F(LZ, LS) \wedge Y)$. We know that $F(LZ, LS) = F(Z, LS) = DZ \wedge LS$, which is L-equivalent to DZ. Thus, the right hand side is $L(DZ \wedge Y) = LF(Z, Y) = F(Z, Y)$, as required.

We now define $\mathcal{G}_L = \{LZ \mid Z \in \mathcal{G}\}$, thereby making L into a stable morphism. If $\mathcal{D}$ is a localizing subcategory of $\mathcal{C}_L$ which contains $\mathcal{G}_L$, then $\{X \in \mathcal{C} \mid LX \in \mathcal{D}\}$ is a localizing subcategory of $\mathcal{C}$ which contains $\mathcal{G}$, hence all of $\mathcal{C}$. It follows that $\mathcal{D} = \mathcal{C}_L$. Thus, with all this structure, $\mathcal{C}_L$ becomes a stable homotopy category.

We still need to show that L preserves (minimal) weak colimits. Suppose that $(\tau_i \colon X_i \longrightarrow X)$ is a weak colimit. This means that $[X, Y] \longrightarrow \varprojlim [X_i, Y]$ is an epimorphism for all Y. As L is a functor, we certainly have compatible maps $(L\tau_i \colon LX_i \longrightarrow LX)$. If $Y \in \mathcal{C}_L$ then $[LX, Y] = [X, Y]$ and $[LX_i, Y] = [X_i, Y]$, so we have an epimorphism $[LX, Y] \longrightarrow \varprojlim [LX_i, Y]$. Thus LX is a weak colimit in $\mathcal{C}_L$ of $\{LX_i\}$.

Now suppose that $(\tau_i \colon X_i \longrightarrow X)$ is a minimal weak colimit, and that H is a homology functor on $\mathcal{C}_L$. Then $H \circ L$ is a homology functor on $\mathcal{C}$, so we must have $H(LX) \simeq \varinjlim H(LX_i)$. Thus $(L\tau_i \colon LX_i \longrightarrow LX)$ is a minimal weak colimit in $\mathcal{C}_L$.

The claims about the inclusion functor J are now easy. We have already seen that $\nu^\#$ and ρ are isomorphisms, and it is easy to see that $\pi^\#$ is also. As $L \simeq 1$ on $\mathcal{C}_L$, we see that $LJ \simeq 1$. □

Proof of Theorem 3.5.2. In this proof we assume that $\mathcal{C}$ is algebraic.

Suppose that L is smashing. Then, for any family $\{X_i\}$ of objects of $\mathcal{C}_L$, the coproduct in $\mathcal{C}$ is already local, and so is the same as the coproduct in $\mathcal{C}_L$. Thus, if $Z \in \mathcal{C}$ is small we have

$$[LZ, \coprod_L X_i] = [Z, \coprod_L X_i] = [Z, \coprod X_i] = \bigoplus [Z, X_i] = \bigoplus [LZ, X_i]$$

so that LZ is small in $\mathcal{C}_L$.

Conversely, suppose that L preserves smallness. Let $\{X_i\}$ be a family of L-local objects. Then, for all $Z \in \Sigma^* \mathcal{G}$, we have isomorphisms

$$[Z, \coprod X_i] = \bigoplus [Z, X_i] = \bigoplus [LZ, X_i] = [LZ, L\left(\coprod X_i\right)] = [Z, L\left(\coprod X_i\right)].$$

Therefore the natural map $\coprod X_i \longrightarrow L(\coprod X_i)$ is an isomorphism, so L is smashing.

Suppose again that L is smashing. Define $K \colon \mathcal{C} \longrightarrow \mathcal{C}$ by $KX = F(LS, X)$. We claim that KX is actually L-local. Indeed, suppose that Z is L-acyclic; then

$$[Z, KX] = [LS \wedge_{\mathcal{C}} Z, X] = [LZ, X] = [0, X] = 0,$$

which implies the claim. We may thus regard K as a functor $\mathcal{C} \to \mathcal{C}_L$, and as such it is easily seen to be right adjoint to J. It follows that J, equipped with the maps

$$\beta\colon S_{\mathcal{C}} \to JS_L$$

$$\nu\colon JU \wedge_{\mathcal{C}} JV \to J(U \wedge_L V)$$

is a lax geometric morphism. It is immediate from the definitions that ν and π are isomorphisms.

We defer to Theorem 4.3.4 the proof that $\mathcal{C}_L$ is a Brown category when $\mathcal{C}$ is a Brown category. □

We next consider the local category obtained by localizing at a set of small objects.

Theorem 3.5.3. *Let $\mathcal{A} \subset \mathcal{C}$ be a thick subcategory of small objects as in Theorem 3.3.5. Let $L = L_{\mathcal{A}}\colon \mathcal{C} \to \mathcal{C}_L$ be the localization functor constructed there, whose category of acyclics is*

$$\mathcal{Z} = \{Y \mid \forall W \in \mathcal{A} \quad [W,Y] = 0\}.$$

Then we can make $\mathcal{C}_L$ into an algebraic stable homotopy category with $\mathcal{G} = \mathcal{A}$, and L into a geometric morphism (but not a stable morphism in general). Moreover, L admits a left adjoint $M = C^f_{\mathcal{A}}$ as well as a right adjoint J. Thus L preserves products, function objects, and sequential limits (as well as the other structure listed in Theorem 3.5.1). The following maps are isomorphisms, where the subscript L indicates structure in $\mathcal{C}_L$.

$$\begin{aligned}
\mu^{\#}\colon LF_{\mathcal{C}}(X,Y) &\simeq F_L(LX,LY)\\
\nu^{\#}\colon JF_L(U,V) &\simeq F_{\mathcal{C}}(JU,JV)\\
\pi^{\#}\colon L(X \wedge_{\mathcal{C}} JU) &\simeq LX \wedge_L U\\
\rho\colon F_{\mathcal{C}}(X,JU) &\simeq JF_L(LX,U)\\
\rho^{\#}\colon LF_{\mathcal{C}}(X,JU) &\simeq F_L(LX,U)
\end{aligned}$$

Moreover, $LM \simeq 1 \simeq LJ\colon \mathcal{C}_L \to \mathcal{C}_L$.

Proof. We know from Theorem 3.5.1 that $\mathcal{C}_L$ is an enriched triangulated category with all cohomology functors representable, and that $L\colon \mathcal{C} \to \mathcal{C}_L$ is a geometric morphism. It follows easily from Theorem 3.3.5 that $L = 1$ on $\mathcal{A}$, so that $\mathcal{A} \subset \mathcal{C}_L$. It also follows that if $Z \in \mathcal{C}$ is such that $[W,Z] = 0$ for all $W \in \mathcal{A}$, then $LZ = 0$. If in addition we have $Z \in \mathcal{C}_L$, then clearly $Z = 0$. It now follows from Theorem 2.3.2 that $\mathcal{C}_L$ is an algebraic stable homotopy category with generators $\mathcal{A}$.

Recall from Theorem 3.3.5 that $LX = F(MS,X)$ and that $MU = MS \wedge U$. It follows that

$$[U,LX] = [U,F(MS,X)] = [MS \wedge U,X] = [MU,X].$$

Thus M is left adjoint to L, which implies that L preserves products and sequential limits.

We know from Theorem 3.3.5 that $ML = M$, and it follows easily that $MU \wedge X = M(U \wedge LX)$. We therefore have

$$\begin{aligned}
[U,LF(X,Y)] &= [MU,F(X,Y)] = [M(U \wedge LX),Y]\\
&= [U \wedge LX,LY] = [U,F(LX,LY)].
\end{aligned}$$

It follows that $LF(X,Y) = F(LX,LY)$, in other words that $\mu^{\#}$ is an isomorphism. It follows in turn that $\rho^{\#}$ is an isomorphism. We saw in Theorem 3.5.1 that $\nu^{\#}$, $\pi^{\#}$ and ρ are isomorphisms, and that $LJ = 1$. We saw in Theorem 3.3.5 that $LM \simeq L$ on $\mathcal{C}$, so that $LM \simeq 1$ on $\mathcal{E} = \mathcal{C}_L$. □

There are many properties one would like $\mathcal{C}_L$ to have that it does not enjoy in general.

Example 3.5.4.

(a) In the homotopy category of spectra $\mathcal{S}$, let L denote localization with respect to MU or with respect to the wedge of all the Morava K-theories $K(n)$ (where $0 \leq n < \infty$). Then there are no nonzero small objects in $\mathcal{S}_L$ [Str].
(b) While every object in thick$\langle L\mathcal{G}\rangle$ is strongly dualizable, there will be other strongly dualizable objects in general. Indeed, any interesting element of the Picard group in the $K(n)$-local category will be strongly dualizable yet not in thick$\langle L_{K(n)}S\rangle$. There are many examples already when $n = 1$ [HMS94].

Note that the subcategory of colocal objects will not, in general, form a stable homotopy category even though coproducts of colocal objects are always colocal. The problem is that colocalization functors would preserve cogenerators, but there is no reason to expect them to preserve generators. This is the main reason that localization functors arise more often than colocalizing functors in stable homotopy theory. See, however, Theorem 9.1.1 for a situation in which the subcategory of colocal objects does form a stable homotopy category.

Next we point out that any localization L, even if it is not smashing, induces correspondences between the full subcategories of $\mathcal{C}_L$ and certain full subcategories of $\mathcal{C}$.

Definition 3.5.5. Suppose that $\mathcal{C}$ is a triangulated category and L is a localization functor. We say that a subcategory $\mathcal{D}$ of $\mathcal{C}$ is *L-replete* if it is full, and whenever $X \to Y$ is an L-equivalence, then $X \in \mathcal{D} \Leftrightarrow Y \in \mathcal{D}$.

Lemma 3.5.6. *Suppose that $\mathcal{C}$ is a triangulated category and L is a localization functor. There is a bijection between replete subcategories of $\mathcal{C}_L$ and L-replete subcategories of $\mathcal{C}$. This correspondence sends thick subcategories to thick subcategories and localizing subcategories to localizing subcategories.*

Proof. If $\mathcal{D}$ is an L-replete full subcategory of $\mathcal{C}$, then define $F(\mathcal{D}) = \mathcal{D} \cap \mathcal{C}_L$. If $\mathcal{E}$ is a full subcategory of $\mathcal{C}_L$, then define $G(\mathcal{E})$ to be the full subcategory of $\mathcal{C}$ with objects $\{X \mid LX \in \mathcal{E}\}$. It is trivial to check that $F \circ G(\mathcal{E}) = \mathcal{E}$, and using L-invariance that $G \circ F(\mathcal{D}) = \mathcal{D}$. We also leave it to the reader to check that if $\mathcal{D} \subseteq \mathcal{C}$ and $\mathcal{E} \subseteq \mathcal{C}_L$ correspond, then $\mathcal{D}$ has the appropriate structure (is thick or localizing) if and only if $\mathcal{E}$ does. The crucial point is that L always preserves coproducts as a functor from $\mathcal{C}$ to $\mathcal{C}_L$. □

We point out that localization functors rarely preserve products (except in the situation of Theorem 3.5.3), so L-replete colocalizing subcategories of $\mathcal{C}$ will not correspond to colocalizing subcategories of $\mathcal{C}_L$.

3.6. **The Bousfield lattice.** In this section we define Bousfield classes and discuss a few of their basic properties; see [Bou79a] and [Rav84] for the proofs and for other useful results.

Definition 3.6.1.

(a) Fix an object X in a stable homotopy category $\mathcal{C}$. We say that an object Y is *X-acyclic* if $X \wedge Y = 0$, and that an object Z is *X-local* if $F(Y, Z) = 0$ for all X-acyclic objects Y.
(b) We define the *Bousfield class* of X (written $\langle X \rangle$) to be the collection of X-local objects. This forms a colocalizing coideal (Definition 1.4.3). We say that two objects X and Y are *Bousfield equivalent* if $\langle X \rangle = \langle Y \rangle$.
(c) The collection of Bousfield classes then defines a partially ordered class under inclusion. Write $\langle X \rangle \amalg \langle Y \rangle$ for $\langle X \amalg Y \rangle$, and $\langle X \rangle \wedge \langle Y \rangle$ for $\langle X \wedge Y \rangle$. (We shall see later that this is well-defined. It is easy to see that $\langle X \rangle \wedge \langle Y \rangle \subseteq \langle X \rangle \cap \langle Y \rangle$, but in general they are not equal—see [Bou79a, Lemma 2.5].)

It is more common to define the Bousfield class of X to be the localizing ideal of X-acyclics. We have chosen to use X-locals instead so that the ordering $\langle X \rangle \geq \langle Y \rangle$ just means that the category of X-locals contains the category of Y-locals.

The partially ordered class of Bousfield classes is contained in an apparently more fundamental lattice called the Bousfield lattice, which we now define.

Let $\mathcal{C}$ be a stable homotopy category. Given two objects X and Y, we write $X \perp Y$ if $F(X, Y) = 0$. More generally, if $\mathcal{D}$ is a class of objects, we write $X \perp \mathcal{D}$ if $X \perp Y$ for all $Y \in \mathcal{D}$, and so on. We define the left and right annihilators of a class $\mathcal{D}$ as follows:

$$^{\perp}\mathcal{D} = \{X \mid X \perp \mathcal{D}\},$$
$$\mathcal{D}^{\perp} = \{X \mid \mathcal{D} \perp X\}.$$

Definition 3.6.2. A class $\mathcal{D} \subseteq \mathcal{C}$ is a *closed localizing ideal* if $\mathcal{D} = {}^{\perp}\mathcal{E}$ for some $\mathcal{E}$. (It is easy to check that such a class is indeed a localizing ideal.) Dually, $\mathcal{D}$ is a *closed colocalizing coideal* if $\mathcal{D} = \mathcal{E}^{\perp}$ for some $\mathcal{E}$.

The purely formal theory of Galois correspondences tells us that the closed localizing ideals form a lattice under inclusion, antiisomorphic to the lattice of closed colocalizing coideals. (We make the lattice operations explicit in the definition below.) Moreover, the smallest closed localizing ideal containing a class $\mathcal{D}$ is ${}^{\perp}(\mathcal{D}^{\perp})$.

Definition 3.6.3. The *Bousfield lattice* of $\mathcal{C}$ is the lattice $\mathcal{L}$ of closed colocalizing coideals. The meet operation is just intersection, and the join operation is

$$\mathcal{D} \amalg \mathcal{E} = ({}^{\perp}\mathcal{D} \cap {}^{\perp}\mathcal{E})^{\perp}.$$

We will refer to an element of the Bousfield lattice as a *generalized Bousfield class*. Note that Bousfield classes are generalized Bousfield classes, because $\langle X \rangle = \{Z \mid X \wedge Z = 0\}^{\perp}$. Note also that if we think of $\mathcal{L}$ as the lattice of closed localizing ideals ordered by reverse inclusion, then the join operation is just intersection.

We do not know whether every generalized Bousfield class is in fact a Bousfield class, although there are a number of generalized Bousfield classes that can only be proved to be Bousfield classes by quite subtle arguments. We do not know whether the collection of Bousfield classes is closed under intersections.

The Bousfield lattice is analogous to the lattice of torsion theories in an Abelian category [Gol86]. However, the lattice of torsion theories has many good properties which we have been unable to prove in our context.

Lemma 3.6.4. *If $L\colon \mathcal{C} \to \mathcal{C}$ is a localization functor, then $\{X \mid LX = 0\}$ is a closed localizing ideal, and $\mathcal{C}_L$ is the corresponding closed colocalizing coideal.*

Proof. See part (b)(iv) of Lemma 3.1.6. □

We do not know in general whether the converse of the above lemma holds, nor do we know whether all localizing ideals are closed. We also do not know whether the closed localizing ideals form a set or a proper class.

We have shown how to associate a Bousfield class to an object X of a stable homotopy category $\mathcal{C}$, and a generalized Bousfield class to a localization functor L. We can also associate a generalized Bousfield class to a homology functor H. Indeed, recall the localizing ideal $\mathcal{D}$ of H-acyclics:

$$\mathcal{D} = \{X : H(X \wedge Y) = 0 \text{ for all } Y\}.$$

Then define $\langle H \rangle = \mathcal{D}^{\perp}$, the closed colocalizing coideal of H-local objects.

Another way to say this, when $\mathcal{C}$ is algebraic, is that $\langle H \rangle$ is equal to the generalized Bousfield class of the localization functor L_H. In this case, there is also a way to associate a homology functor to an object X. Recall that we defined the category $\Lambda(X)$ in Definition 2.3.7.

Definition 3.6.5. Let $\mathcal{C}$ be an algebraic stable homotopy category. Write

$$\widehat{\pi}_0(X) = \varinjlim{}_{\Lambda(X)}[S, X_\alpha].$$

This is a homology functor by Corollary 2.3.11. If $\mathcal{C}$ is unital algebraic, then $\widehat{\pi}_0(X) = \pi_0(X)$. Given an object $X \in \mathcal{C}$, we define

$$H_X(Y) = \widehat{\pi}_0(X \wedge Y).$$

This is again a homology functor. There is an obvious natural map $H_X(Y) = H_Y(X) \to \pi_0(X \wedge Y) = X_0 Y$, which is an isomorphism when $\mathcal{C}$ is unital algebraic.

Thus, in an algebraic stable homotopy category, there are two generalized Bousfield classes associated to an object X. The following lemma shows they are in fact equal.

Lemma 3.6.6. *Let X be an object of an algebraic stable homotopy category $\mathcal{C}$. Then for any $Y \in \mathcal{C}$ we have $X \wedge Y = 0$ if and only if Y is H_X-acyclic, and thus $\langle X \rangle = \langle H_X \rangle$. The category of such Y is a closed localizing subcategory.*

Proof. If Y is X-acyclic, then $X \wedge Y = 0$, so $H_X(Y \wedge Z) = \widehat{\pi}_0(X \wedge Y \wedge Z) = 0$ for all Z. Thus Y is H_X-acyclic. Conversely, suppose Y is H_X-acyclic. By Lemma 4.1.2, we find that $\pi_0(X \wedge Y \wedge Z) = 0$ for all small Z. By Spanier-Whitehead duality, we find that $[Z, X \wedge Y] = 0$ for all small Z, and in particular for all $Z \in \mathcal{G}$. Thus, by Lemma 1.4.5, $X \wedge Y = 0$ and so Y is X-acyclic. The localizing category of H_X-acyclics is closed by Theorem 3.2.2 and Lemma 3.6.4. The categories $\langle X \rangle$ and $\langle H_X \rangle$ are by definition the right annihilators of the X-acyclics and the H_X-acyclics, so they are the same. □

Because of this lemma, we denote the localization functor L_{H_X} simply by L_X.

An annoying feature of this lemma is that we have to define the acyclics of H_X by requiring that $H_X(Y \wedge Z) = 0$ for all Z. It is sometimes useful to avoid this.

Definition 3.6.7. An object X in an algebraic stable homotopy category $\mathcal{C}$ is *monoidal* if $H_{X*}(Y) = 0$ implies $H_{X*}(Y \wedge Z) = 0$ for all Z. (Note the grading on H_X.)

We write $L_1X = L_1S \wedge X$ for the cofiber of the natural map $C_1X \to X$; note that this is not the same as L_0X. An argument similar to the above shows that C_1 is right adjoint to the inclusion of $\mathcal{C}^C \cap \mathcal{C}^{C'}$ in $\mathcal{C}$, so it is a colocalization functor with $\mathcal{C}^{C_1} = \mathcal{C}^C \cap \mathcal{C}^{C'}$. It follows that L_1 is a (smashing) localization with $\mathcal{C}_{L_1} = \mathcal{C}_L \amalg \mathcal{C}_{L'}$ as required.

It follows from Lemma 3.8.1 that the operations $\wedge$ and $\cap$ agree on $\mathfrak{L}_s$, and $\wedge$ clearly distributes over $\amalg$, so $\mathfrak{L}_s$ is distributive.

Now consider a countable family of smashing localizations $\{L_k\}$, whose meet we wish to construct. We may replace L_k by the meet of $L_0, \dots, L_k$ and thus assume that we have a descending sequence

$$\langle L_0S\rangle \geq \langle L_1S\rangle \geq \langle L_2S\rangle \geq \dots$$

By Lemma 3.1.5, there is unique morphism $L_k \to L_{k+1}$ for each k. We define $L_\infty S$ to be the sequential limit of the objects L_kS, so there is an obvious map $i_\infty = (S \xrightarrow{i_0} L_0S \to L_\infty S)$. Write $L_\infty X = L_\infty S \wedge X$, which is clearly a functor of X. We also see that $L_\infty X$ is the sequential colimit of the sequence L_kX.

For each k, we have a map of cofiber sequences

$$\begin{array}{ccccc} C_kS & \longrightarrow & S & \longrightarrow & L_kS \\ \downarrow & & \downarrow{=} & & \downarrow \\ C_{k+1}S & \longrightarrow & S & \longrightarrow & L_{k+1}S \end{array}$$

We would like to take the sequential colimit of these maps to get a cofiber sequence, but the sequential colimit is not always exact. Nonetheless, we get a commutative diagram

$$\begin{array}{ccccc} \coprod_k C_kS & \longrightarrow & \coprod_k S & \xrightarrow{\coprod_k i_k} & \coprod_k L_kS \\ & & {\scriptstyle f}\downarrow & & {\scriptstyle g}\downarrow \\ \coprod_k C_kS & \longrightarrow & \coprod_k S & \xrightarrow{\coprod_k i_k} & \coprod_k L_kS \\ & & {\scriptstyle p}\downarrow & & {\scriptstyle q}\downarrow \\ & & S & & L_\infty S \end{array}$$

Here the maps f and g are the usual ones, whose cofibers are the respective sequential colimits. By the analysis in Lemma 2.2.6, we see that $pj_0 = 1$, where $j_0 \colon S \to \coprod_k S$ includes the first factor. The 3×3 Lemma A.1.2 now gives us a diagram

$$\begin{array}{ccccc} \coprod_k C_kS & \longrightarrow & \coprod_k S & \xrightarrow{\coprod_k i_k} & \coprod_k L_kS \\ \downarrow & & {\scriptstyle p}\downarrow & & \downarrow{\scriptstyle q} \\ C_\infty S & \longrightarrow & S & \xrightarrow[i]{} & L_\infty S \end{array}$$

Here the rows are exact, and $C_\infty S$ is a the cofiber of some self-map of $\coprod_k C_kS$ and i is some map $S \to L_\infty S$. In fact, we have

$$i = ipj_0 = q\left(\coprod i_k\right) j_0 = qi_0 = i_\infty.$$

We can restate this as follows: the fiber $C_\infty S$ of $i_\infty : S \longrightarrow L_\infty S$ lies in the localizing subcategory generated by the objects $C_k S$.

Note that $C_k S \wedge L_m S = 0$ for $m \geq k$, so $C_k S \wedge L_\infty S = \varinjlim_m C_k S \wedge L_m S = 0$. It follows that $L_\infty S \wedge C_\infty S = 0$.

We write $C_\infty X = C_\infty S \wedge X$, and $\mathcal{D} = \{X \mid C_\infty X = 0\}$. By the above, L_∞ can be considered as a functor $\mathcal{C} \longrightarrow \mathcal{D}$. We claim that $\mathcal{D} = \bigcap_k \mathcal{C}_{L_k}$. Indeed, suppose that $X \in \bigcap_k \mathcal{C}_{L_k}$, so that $C_k S \wedge X = 0$ for all $k < \infty$. As $C_\infty S$ lies in the localizing subcategory generated by the $C_k S$, we see that $C_\infty S \wedge X = 0$ and thus $X \in \mathcal{D}$. Conversely, suppose that $X \in \mathcal{D}$. Then $X = L_\infty X = \varinjlim_m L_m X$. We may start the colimit at the kth stage; as $C_k L_m X = 0$ for $k \leq m < \infty$, we conclude that $C_k X = C_k L_\infty X = 0$. Thus $X \in \bigcap_k \mathcal{C}_{L_k}$.

Next, we claim that $[C_\infty X, L_\infty Y] = 0$ for all X and Y. Indeed, we have just seen that $L_\infty Y$ is L_k-local for all k, so $[C_k X, L_\infty Y] = 0$ for all k. As $C_\infty X \in \operatorname{loc}\langle C_k X \mid k \geq 0\rangle$, we see that $[C_\infty X, L_\infty Y] = 0$.

Next, we claim that L_∞ is left adjoint to the inclusion of $\mathcal{D}$ in $\mathcal{C}$. Indeed, suppose that $Y \in \mathcal{D}$, so $Y = L_\infty Y$, so $[C_\infty X, Y] = 0$ for all Y. Thus the cofibration $C_\infty X \longrightarrow X \longrightarrow L_\infty X$ shows that $[X, Y] = [L_\infty X, Y]$ as required. This implies that L_∞ is a smashing localization with $\mathcal{C}_{L_\infty} = \bigcap_k \mathcal{C}_{L_k}$ as required. □

The sublattice $\mathcal{L}_s$ of the Bousfield lattice is not closed under countable joins. Indeed, in the category of spectra, the join of the smashing Bousfield classes $\langle E(n)\rangle$ is the harmonic Bousfield class, which is not smashing.

The situation is even better for finite localizations, as expressed by the following result.

Proposition 3.8.3. *Let $\mathcal{C}$ be an algebraic stable homotopy category, and let $\mathcal{L}_f$ be the collection of Bousfield classes of the form $\langle L^f_{\mathcal{A}}\rangle$, where $L^f_{\mathcal{A}}$ is a finite localization functor. Then $\mathcal{L}_f$ is closed under finite joins and arbitrary meets. It is a distributive lattice, antiisomorphic to the lattice of $\mathcal{G}$-ideals of small objects (which is a set rather than a proper class).*

Proof. Let $\mathcal{F}$ be the category of small objects in $\mathcal{C}$, and $\mathcal{L}'_f$ the collection of $\mathcal{G}$-ideals of $\mathcal{F}$. This is clearly a lattice with arbitrary meets (given by intersection) and joins (given by taking the $\mathcal{G}$-ideal generated by the union). As $\mathcal{C}$ is algebraic, there is only a set of small objects up to isomorphism, so $\mathcal{L}'_f$ is a set. Suppose that $\mathcal{A} \in \mathcal{L}'_f$. Because $\mathcal{A}$ is a $\mathcal{G}$-ideal, it is easy to see that $\operatorname{loc}\langle\mathcal{A}\rangle = \operatorname{locid}\langle\mathcal{A}\rangle$. Theorem 3.3.3 gives a smashing localization functor $L^f_{\mathcal{A}}$ with

$$\langle L^f_{\mathcal{A}} S\rangle = \mathcal{C}_{L^f_{\mathcal{A}}} = \operatorname{loc}\langle\mathcal{A}\rangle^{\perp}$$

and

$$\ker(L^f_{\mathcal{A}}) = \operatorname{loc}\langle\mathcal{A}\rangle.$$

Moreover, it tells us that $\operatorname{loc}\langle\mathcal{A}\rangle \cap \mathcal{F} = \mathcal{A}$. It follows directly that the map $j \colon \mathcal{A} \mapsto \langle L^f_{\mathcal{A}}\rangle$ is an order-reversing map from $\mathcal{L}'_f$ to the Bousfield lattice $\mathcal{L}$. If we view $\mathcal{L}$ as the opposite of the lattice of closed localizing subcategories, then j becomes the map $\mathcal{A} \mapsto \operatorname{loc}\langle\mathcal{A}\rangle$, and this makes it easy to see that j sends arbitrary joins to meets. Also, as $\mathcal{A} = \operatorname{loc}\langle\mathcal{A}\rangle \cap \mathcal{F}$, we see that j is a monomorphism. The image is by definition $\mathcal{L}_f$, so $\mathcal{L}_f$ is closed under arbitrary meets.

Now consider two elements $\mathcal{A}_0, \mathcal{A}_1$ of $\mathcal{L}'_f$, and write $\mathcal{A}_2 = \mathcal{A}_0 \cap \mathcal{A}_1$. We claim that $j(\mathcal{A}_2)$ is the join of $j(\mathcal{A}_0)$ and $j(\mathcal{A}_1)$ in the Bousfield lattice. It will suffice to show that $\mathrm{loc}\langle \mathcal{A}_2\rangle = \mathrm{loc}\langle \mathcal{A}_0\rangle \cap \mathrm{loc}\langle \mathcal{A}_1\rangle$. It is immediate that $\mathrm{loc}\langle \mathcal{A}_2\rangle \subseteq \mathrm{loc}\langle \mathcal{A}_0\rangle \cap \mathrm{loc}\langle \mathcal{A}_1\rangle$. For the converse, we first recall that small objects in an algebraic stable homotopy category are $\mathcal{G}$-finite, which implies that $W \wedge X \in \mathcal{A}_2$ whenever $W \in \mathcal{A}_0$ and $X \in \mathcal{A}_1$ (or in other words, $\mathcal{A}_0 \wedge \mathcal{A}_1 \subseteq \mathcal{A}_2$). By considering $\{Z \mid Z \wedge \mathcal{A}_1 \subseteq \mathrm{loc}\langle \mathcal{A}_2\rangle\}$, we conclude that $\mathrm{loc}\langle \mathcal{A}_0\rangle \wedge \mathcal{A}_1 \subseteq \mathrm{loc}\langle \mathcal{A}_2\rangle$. A similar argument then shows that $\mathrm{loc}\langle \mathcal{A}_0\rangle \wedge \mathrm{loc}\langle \mathcal{A}_1\rangle \subseteq \mathrm{loc}\langle \mathcal{A}_2\rangle$. In particular (if we write C_i for $C^f_{\mathcal{A}_i}$) we have $C_0 S \wedge C_1 S \in \mathrm{loc}\langle \mathcal{A}_2\rangle$. If $X \in \mathrm{loc}\langle \mathcal{A}_0\rangle \cap \mathrm{loc}\langle \mathcal{A}_1\rangle$ then $X = C_0 X = C_1 X$ so $X = X \wedge (C_0 S \wedge C_1 S)$, so $X \in \mathrm{loc}\langle \mathcal{A}_2\rangle$ as required.

This shows that $\mathfrak{L}_f$ is closed under the lattice operations in $\mathfrak{L}_s$, so it is a distributive lattice. □

4. Brown representability

In this section we discuss the representability of homology functors and related issues. Throughout this section, $\mathcal{C}$ will be an algebraic stable homotopy category.

4.1. Brown categories.

We begin with the definition of a representable homology functor and of a Brown category. The first part of the definition below appeared as Definition 3.6.5, but we repeat it here for convenience.

Definition 4.1.1. Write

$$\widehat{\pi}_0(X) = \varinjlim_{\Lambda(X)} [S, X_\alpha].$$

This is a homology functor by Corollary 2.3.11. If $\mathcal{C}$ is unital algebraic, then $\widehat{\pi}_0(X) = \pi_0(X)$. Given an object $X \in \mathcal{C}$, we define

$$H_X(Y) = \widehat{\pi}_0(X \wedge Y).$$

This is again a homology functor. There is an obvious natural map $H_X(Y) = H_Y(X) \longrightarrow \pi_0(X \wedge Y)$.

A homology functor H on $\mathcal{C}$ is *representable* if there is an object Y of $\mathcal{C}$ and an isomorphism of homology functors $H_Y \simeq H$. Note that this definition is inconsistent with the usual categorical terminology, in which a covariant functor is said to be representable if and only if it is equivalent to a functor of the form $[Y, -]$. Nonetheless, it is close to the standard usage in stable homotopy theory.

Lemma 4.1.2. *If X is small then the natural map $H_X(Y) \longrightarrow \pi_0(X \wedge Y)$ is an isomorphism.*

Proof. Recall from Theorem 2.1.3 that the subcategory of small objects is closed under smash products and the duality functor $DX = F(X, S)$.

Suppose that X is small. Then $\pi_0(X \wedge Y) = [DX, Y]$ is a homology functor of Y, by the smallness of DX. When Y is also small, then so is $X \wedge Y$, so $H_X(Y) = \pi_0(X \wedge Y)$. Thus the natural map $H_X(Y) \longrightarrow \pi_0(X \wedge Y)$ is a map of homology functors which is an isomorphism for small Y, and thus for all Y. □

Lemma 4.1.3. *There is a natural isomorphism*

$$H_X(Y) \simeq \varinjlim_{\Lambda(Y)} \pi_0(X \wedge Y_\alpha).$$

Proof. Both sides are homology functors of Y (by Corollary 2.3.11). They are isomorphic for small Y by Lemma 4.1.2 and the fact that $H_X Y = H_Y X$. Thus, they are isomorphic for all Y. □

We can now define a Brown category.

Definition 4.1.4. A *Brown category* is an algebraic stable homotopy category $\mathcal{C}$ such that every homology functor is representable and every natural transformation $H_X \longrightarrow H_Y$ of homology functors is induced by a (typically nonunique) map $X \longrightarrow Y$.

Naturally one would like to be able to tell when an algebraic stable homotopy category is a Brown category. Neeman shows in [Nee95] that the derived category of modules over $\mathbf{C}[x, y]$ is not a Brown category, although it is a monogenic stable homotopy category, so it seems that the Brown condition is a genuinely subtle one.

Recall, from Definition 2.3.3, that $c(\mathcal{C})$ denotes the (necessarily infinite) cardinality of the disjoint union of the sets $[W, Z]_n$ for all $W, Z \in \mathcal{G}$ and $n \in \mathbf{Z}$.

Theorem 4.1.5. *If $\mathcal{C}$ is an algebraic stable homotopy category with $c(\mathcal{C}) = \aleph_0$, then $\mathcal{C}$ is a Brown category.*

Proof. Suppose that $\mathcal{C}$ is algebraic and $c(\mathcal{C}) = \aleph_0$. Let $H\colon \mathcal{F}^{\mathrm{op}} \longrightarrow \mathrm{Ab}$ be an exact functor. We claim that this can be extended to give a cohomology functor defined on all of $\mathcal{C}$. The proof of this fact is complicated, but it is also the same, *mutatis mutandis*, as the proof in [Mar83, Chapter 4] (which in turn follows [Ada71]). We give a brief outline. First, we define a functor $\widehat{H}\colon \mathcal{C}^{\mathrm{op}} \longrightarrow \mathrm{Ab}$ by

$$\widehat{H}(Y) = \varprojlim{}_{\Lambda(Y)} H(Y_\alpha).$$

This converts coproducts to products, minimal weak colimits to limits, and with considerable work one can show that is exact on a restricted class of cofibrations. This uses the that fact a countable filtered diagram of nonempty sets and surjections has nonempty inverse limit; it is here that the countability hypothesis is used. (The analogous statement is false for an inverse system indexed by the first uncountable ordinal.) In any case, this restricted exactness suffices to carry through most of the proof of representability as in Theorem 2.3.2. This gives an object X and a natural map $[Y, X] \longrightarrow \widehat{H}(Y)$ which is an equivalence when $Y \in \Sigma^*\mathcal{G}$ and thus when $Y \in \mathcal{F}$. Of course, when $Y \in \mathcal{F}$ we also have $\widehat{H}(Y) = H(Y)$. We may therefore take $[-, X]$ as an extension of H to all of $\mathcal{C}$.

Now suppose that we have two contravariant exact functors $H', H''\colon \mathcal{F}^{\mathrm{op}} \longrightarrow \mathrm{Ab}$ and a natural map $f\colon H' \longrightarrow H''$. Choose objects X', X'' representing H' and H'' as above. Margolis also proves that f arises from a map $g\colon X' \longrightarrow X''$.

Now let $H\colon \mathcal{C} \longrightarrow \mathrm{Ab}$ be a homology functor. By applying the above to $H' = H \circ D$ (where D is the Spanier-Whitehead duality functor), we obtain an object X and an equivalence $H(Z) = H'(DZ) = [DZ, X] = X_0 Z = H_X(Z)$ for small objects Z. Using Corollary 2.3.11, we see that $H(Z) = H_X(Z)$ for all Z, so that H is representable. The previous paragraph shows that maps are also representable. □

In a Brown category, every natural transformation of homology functors is induced by a map of the representing objects, but there may be more than one such map inducing the same natural transformation.

To investigate this nonuniqueness, we adopt a somewhat abstract approach. Suppose that $\mathcal{C}$ is an algebraic stable homotopy category, and let $\mathcal{F}$ be the full subcategory of small objects. We write $\mathcal{F}_\bullet$ and $\mathcal{F}^\bullet$ for the categories of covariant and contravariant exact functors $\mathcal{F} \longrightarrow \mathrm{Ab}$. Similarly, we write $\mathcal{C}_\bullet$ and $\mathcal{C}^\bullet$ for the categories of homology and cohomology functors from $\mathcal{C}$ to Ab. Composition with the Spanier-Whitehead duality functor D gives an equivalence $\mathcal{F}_\bullet \simeq \mathcal{F}^\bullet$, which we also call D. Restriction gives functors $U_\bullet\colon \mathcal{C}_\bullet \longrightarrow \mathcal{F}_\bullet$ and $U^\bullet\colon \mathcal{C}^\bullet \longrightarrow \mathcal{F}^\bullet$. We also have functors $V_\bullet\colon \mathcal{C} \longrightarrow \mathcal{C}_\bullet$ and $V^\bullet\colon \mathcal{C} \longrightarrow \mathcal{C}^\bullet$, sending X to H_X and X^0 respectively.

Finally, we define a functor $L\colon \mathcal{F}_\bullet \longrightarrow \mathcal{C}_\bullet$ by $LH = \widehat{H}_{\mathcal{F}}$. Recall from Section 2.3 that $LH(X)$ is the colimit of $H(Y)$ for all small objects Y over X.

Note that all of these functors except $V_\bullet$ and L are defined on any stable homotopy category.

Definition 4.1.6. A map $f\colon X \longrightarrow Y$ in a stable homotopy category $\mathcal{C}$ is *phantom* if, for all small objects Z and maps $g\colon Z \longrightarrow X$, the composite $f \circ g$ is trivial. Equivalently, f is phantom if and only if $U^\bullet V^\bullet f = 0$.

Remark 4.1.7. Margolis [Mar83] calls these maps f-phantom maps; his definition of phantoms is slightly different.

The phantom maps clearly form an ideal, in the following sense: if $f, g\colon X \longrightarrow Y$ are phantom, and $u\colon W \longrightarrow X$ and $v\colon Y \longrightarrow Z$ are arbitrary, then $f+g$, fu and vf are phantom. We denote the subgroup of $[X, Y]$ consisting of phantom maps by $\mathcal{P}(X, Y)$.

The above functors fit into a commutative diagram as follows:

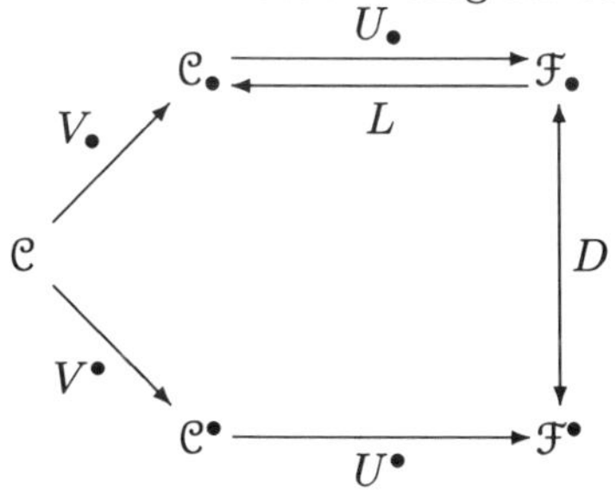

The following theorem is mainly a compendium of results and definitions that we have already seen.

Theorem 4.1.8. *Suppose that $\mathcal{C}$ is a Brown category. Then the functors D, $U_\bullet$ and $V^\bullet$ are equivalences, and L is inverse to $U_\bullet$. The functors $U^\bullet$ and $V_\bullet$ are full and essentially surjective, and reflect isomorphisms. The kernel of $V_\bullet$ (or $U^\bullet$) is a square-zero ideal. To make some of these claims more explicit*:

(a) *Any homology functor H on $\mathcal{F}$ extends to a homology functor LH on $\mathcal{C}$, unique up to canonical isomorphism.*

(b) *The group of natural maps $f\colon H_X \longrightarrow H_Y$ is isomorphic to $[X, Y]/\mathcal{P}(X, Y)$. A map $f\colon X \longrightarrow Y$ is an isomorphism if and only if the induced map $f_0\colon H_X \longrightarrow H_Y$ is an isomorphism.*

(c) *Any homology functor H on $\mathcal{C}$ is equivalent to H_X for some $X \in \mathcal{C}$. The representing object X is unique up to isomorphism, but the isomorphism is only canonical up to the addition of a phantom map.*

(d) *The composite of any two phantom maps is zero.*

Proof. The functor $V^\bullet$ is an equivalence in any stable homotopy category. Indeed, Yoneda's lemma tells us $V^\bullet$ is full and faithful, and since every cohomology functor is representable, it is essentially surjective. Thus $V^\bullet$ is an equivalence. As $D\colon \mathcal{F}^{op} \longrightarrow \mathcal{F}$ satisfies $D^2 \simeq 1$, we see that $D\colon \mathcal{F}_\bullet \longrightarrow \mathcal{F}^\bullet$ is an equivalence in any stable homotopy category. It follows easily from Propositions 2.3.9 and 2.3.1 that $U_\bullet$ and L are inverse equivalences in any algebraic stable homotopy category.

In view of the above and our commutative diagram of functors, our claims about $U^\bullet$ are equivalent to the corresponding claims about $V_\bullet$. Now $V_\bullet$ is full and essentially surjective by the definition of a Brown category. It is also clear that the kernel of $V_\bullet$ is the same as that of $U^\bullet V^\bullet$, which is the ideal of phantoms. In other words, a map $f\colon X \to Y$ induces the zero map $H_X \to H_Y$ if and only if f is phantom.

We next observe that $V_\bullet$ reflects isomorphisms. In other words, if a map $f\colon X \to Y$ is such that $f_0\colon H_X \to H_Y$ is an isomorphism, then f is an isomorphism. Indeed, for $Z \in \Sigma^*\mathcal{G}$ we observe that f gives an isomorphism between $[Z, X] = H_X DZ$ and $[Z, Y] = H_Y DZ$, so the claim follows by Lemma 1.4.5.

This leaves only the claim that the composite of two phantom maps is zero. We will prove this as Theorem 4.2.5 below. □

4.2. **Minimal weak colimits.** The goal of this section is to show that an algebraic stable homotopy category is a Brown category if and only if all filtered minimal weak colimits of small objects exist. We use this, following Christensen [Chr], to show that the composite of any two phantom maps in a Brown category is trivial.

We begin by restating and proving Proposition 2.2.2, which gives an easier criterion for a weak colimit to be minimal in a Brown category.

Proposition 4.2.1. *Let $\mathcal{C}$ be a Brown category. Suppose that $\mathcal{I}$ is a small filtered category, $i \mapsto X_i$ is a functor from $\mathcal{I}$ to $\mathcal{C}$, and $(\tau_i\colon X_i \to X)$ is a weak colimit. Then X is the minimal weak colimit if and only if the induced map*

$$\varinjlim [Z, X_i]_* \to [Z, X]_*$$

is an isomorphism for all $Z \in \mathcal{G}$.

Proof. Suppose that X is the minimal weak colimit, and that $Z \in \mathcal{G}$. Then Z is small, so $[\Sigma^k Z, -]$ is a homology theory. It follows immediately that

$$\varinjlim [Z, X_i]_* \to [Z, X]_*$$

is an isomorphism.

Conversely, suppose that the above map is an isomorphism for all $Z \in \mathcal{G}$. We need to prove that $\varinjlim_i H(X_i) = H(X)$ for all homology functors H on $\mathcal{C}$. As $\mathcal{C}$ is a Brown category, we need only show that

$$H_U(X) = \varinjlim_i H_U(X_i)$$

for all $U \in \mathcal{C}$. The left hand side is the same as $H_X(U)$, and the right hand side is $\varinjlim_i H_{X_i}(U)$. These are both homology functors of U (using the exactness of filtered colimits). By hypothesis, they agree when $U \in \mathcal{G}$, so they agree for all U. □

We also recall the following lemma, whose proof we have given in Remark 2.3.18.

Lemma 4.2.2. *Suppose that $\mathcal{C}$ is an algebraic stable homotopy category, and X is an object of $\mathcal{C}$. If the diagram $\Lambda(X)$ of small objects over X has a minimal weak colimit Y, then $X \simeq Y$.* □

We can now prove the main theorem of this section.

Theorem 4.2.3. *Let $\mathcal{C}$ be an algebraic stable homotopy category, and $\mathcal{F} \subset \mathcal{C}$ the subcategory of small objects. Then $\mathcal{C}$ is a Brown category if and only if any functor $i \mapsto X_i$ from a small filtered category $\mathcal{I}$ to $\mathcal{F}$ has a minimal weak colimit.*

Proof. First suppose that $\mathcal{C}$ is a Brown category. Just as in [Mar83, Theorem 5.13], we define a functor $H\colon \mathcal{C} \to \mathrm{Ab}$ by

$$H(U) = \varinjlim_{\mathcal{I}} H_{X_i}(U) = \varinjlim_{\mathcal{I}} \pi_0(U \wedge X_i).$$

Here the latter isomorphism holds since X_i is small. Then H is a homology functor, since filtered colimits of exact sequences of Abelian groups are exact. Let X be the representing object for H. There are maps $X_i \xrightarrow{f_i} X$ induced by including one factor into the colimit. This map is uniquely determined up to a phantom map, but since X_i is small, there are no phantom maps out of X_i. By the uniqueness, the maps f_i are compatible. That is, given a map $s\colon i \longrightarrow j$ in $\mathcal{I}$, we have $f_j \circ X_s = f_i$.

Now if we have a compatible family of maps $g_i\colon X_i \longrightarrow Y$, we get a natural transformation of homology functors $H_X \longrightarrow H_Y$ since H_X is just the colimit of the H_{X_i}. This map is induced by a map $X \xrightarrow{h} Y$, which is unique up to a phantom map. Again, since the X_i are small, we must have $hf_i = g_i$. Thus X is a weak colimit.

To see that X is the minimal weak colimit, suppose that $Z \in \Sigma^* \mathcal{G}$. It suffices to show that $\varinjlim [Z, X_i] = [Z, X]$ by Proposition 4.2.1. But

$$\varinjlim [Z, X_i] = \varinjlim H_{X_i}(DZ) = H_X(DZ) = [Z, X]$$

as required.

Now suppose that $\mathcal{C}$ is algebraic, and every filtered diagram of small objects has a minimal weak colimit. To show that $\mathcal{C}$ is a Brown category, we will work with contravariant exact functors on $\mathcal{F}$ (the class of small objects) rather than homology functors. This is equivalent to working with homology functors by the argument of Theorem 4.1.5 or Theorem 4.1.8. So, suppose that we have a contravariant exact functor $H\colon \mathcal{F}^{\mathrm{op}} \longrightarrow \mathrm{Ab}$. We will show that H is representable. Define a category $\mathcal{I}_H$ whose objects are pairs (Z, z) where $Z \in \mathcal{F}$ and $z \in H(Z)$. A map $(Z, z) \longrightarrow (Z', z')$ is a map $f\colon Z \longrightarrow Z'$ such that $H(f)(z') = z$. Since the class of small objects is essentially small, and $H(Z)$ is a set for each Z, $\mathcal{I}_H$ is also essentially small. We claim that $\mathcal{I}_H$ is filtered. Indeed, given two objects (Z, z) and (Z', z'), we have the obvious morphisms $(Z, z) \longrightarrow (Z \amalg Z', (z, z')) \longleftarrow (Z', z')$. Also, if we have two morphisms $f, g\colon (Z, z) \longrightarrow (Z', z')$, we let $h\colon Z' \longrightarrow W$ denote the cofiber of $f - g$. Since $H(f - g)(z') = 0$, there is a $w \in H(W)$ such that $H(h)(w) = z'$, so a morphism $h\colon (Z', z') \longrightarrow (W, w)$ coequalizing f and g.

We have an evident functor $\mathcal{I}_H \longrightarrow \mathcal{F}$ that takes (Z, z) to Z. Let X denote the minimal weak colimit of this functor. Then we have compatible maps $i_{(Z,z)}\colon Z \longrightarrow X$ for all objects (Z, z) of $\mathcal{I}_H$. We will construct a natural equivalence $X^0 \longrightarrow H$ on $\mathcal{F}$. To do so, recall that we can extend H to $\mathcal{C}$ by defining

$$\widehat{H}(Y) = \varprojlim{}_{\Lambda(Y)} H(Y_\alpha).$$

(This is not generally a cohomology functor on $\mathcal{C}$.) We define a canonical class $x \in \widehat{H}(X)$ as follows. Since X is a minimal weak colimit, a map $W \xrightarrow{g} X$ in $\Lambda(X)$ factors as $g = i_{(Z,z)} \circ g'$ for some object (Z, z) of $\mathcal{I}_H$ and some map $g'\colon W \longrightarrow Z$. We can then define $x_{(W,g)} = H(g')(z) \in H(W)$. Then $x_{(W,g)}$ is well-defined and defines a class $x \in \widehat{H}(X)$.

We then have a natural transformation $X^0 \longrightarrow \widehat{H}$ that takes a map $W \xrightarrow{g} X$ to $\widehat{H}(g)(x)$. To see that this natural transformation is always surjective, suppose $w \in \widehat{H}(W)$. Then for each $(Z_\alpha \longrightarrow W) \in \Lambda(W)$, we have a class $w_\alpha \in H(Z_\alpha)$, and the w_α are compatible. This gives us compatible maps $i_{(Z_\alpha, w_\alpha)}\colon Z \longrightarrow X$. Since W is a weak colimit of $\Lambda(W)$, by Lemma 4.2.2, there is a map $W \xrightarrow{g} X$ extending the $i_{(Z_\alpha, w_\alpha)}$. It is then easy to see that $\widehat{H}(g)(x) = w$.

Now suppose that W is small, and we have a map $g\colon W \longrightarrow X$ such that $\widehat{H}(g)(x) = 0$. Then there is an object (Z, z) of $\mathcal{I}_H$ and a map $g'\colon W \longrightarrow Z$ such that $g = i_{(Z,z)} \circ g'$. In particular, $H(g')(z) = 0$, so g' is a morphism in the category $\mathcal{I}_H$ from $(W, 0)$ to (Z, z). Thus we have $i_{(W,0)} = i_{(Z,z)} \circ g'$. Thus $g = i_{(W,0)}$. On the other hand, the zero map is a morphism in $\mathcal{I}_H$ from $(W, 0)$ to itself, and from this it follows that $i_{(W,0)}$ is trivial.

Therefore, the natural transformation $X^0 \longrightarrow \widehat{H}$ is an isomorphism on $\mathcal{F}$. Thus every exact contravariant functor on $\mathcal{F}$ is representable. To complete the proof that $\mathcal{C}$ is a Brown category, we must show that any natural transformation of contravariant exact functors on $\mathcal{F}$ is representable. So suppose that H and K are exact contravariant functors on $\mathcal{F}$, and $\tau\colon H \longrightarrow K$ is a natural transformation. We then get a functor $\mathcal{I}_H \longrightarrow \mathcal{I}_K$ that takes (Z, z) to $(Z, \tau(z))$. If we let X denote the minimal weak colimit of $\mathcal{I}_H \longrightarrow \mathcal{F}$, and Y denote the minimal weak colimit of $\mathcal{I}_K \longrightarrow \mathcal{F}$, then we get a (non-unique) induced map $X \longrightarrow Y$. This is the required representation of τ. □

Theorem 4.2.4. *Suppose that $\mathcal{C}$ is a Brown category. Then any object $X \in \mathcal{C}$ is the minimal weak colimit of the diagram $\Lambda(X)$ (defined in Definition 2.3.7).*

Proof. This is immediate from Lemma 4.2.2 and Theorem 4.2.3. □

The following result is well-known in the case of spectra with countable homotopy groups, or with a slightly different notion of phantom maps defined in terms of finite-dimensional spectra rather than finite spectra. However, it seems less well-known in full generality. It is probably due to Boardman, but we learned how to prove it from Dan Christensen [Chr]. It is also proved in [Nee95].

Theorem 4.2.5. *In a Brown category $\mathcal{C}$, the composite of two phantom maps is trivial.*

Proof. We have seen that X is the minimal weak colimit of $\Lambda(X)$. Let us write X_α for a generic object of $\Lambda(X)$, and $u\colon X_\alpha \longrightarrow X_\beta$ for a generic morphism.

As in the proof of Proposition 2.2.4, we have a non-minimal weak colimit C, defined as the cofiber in a sequence

$$\coprod_u X_\alpha \to \coprod_\alpha X_\alpha \to C.$$

Let F denote the fiber of the evident map $\coprod_\alpha X_\alpha \longrightarrow X$. Because X is the minimal weak colimit, we get a split monomorphism $a\colon X \longrightarrow C$ of objects under $\coprod_\alpha X_\alpha$ (using part (c) of 2.2.4). Applying the octahedral axiom to the morphisms

$\coprod_\alpha X_\alpha \to X \xrightarrow{a} C$, we get a diagram as follows.

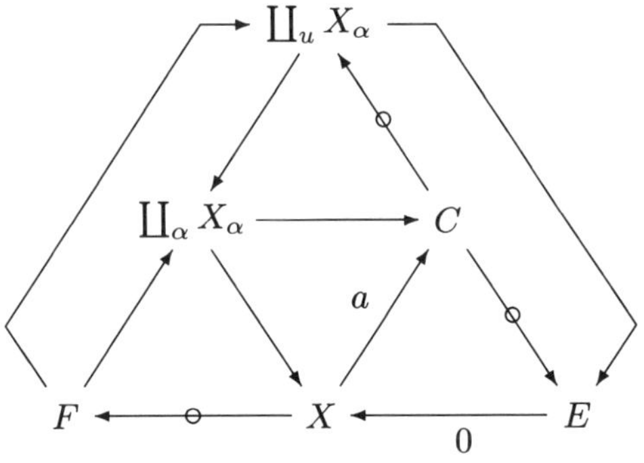

Because $X \to C$ is a split monomorphism, we see that $E \to X$ is zero, so $E \to F$ is zero, so $F \to \coprod_u X_\alpha$ is a split monomorphism, so F is a retract of $\coprod_u X_\alpha$. As $\coprod_u X_\alpha$ is a coproduct of finite objects, we see that there can be no phantom maps out of F.

We have just put an arbitrary object X in a cofiber sequence

$$F \xrightarrow{u} \coprod_\alpha X_\alpha \xrightarrow{v} X \xrightarrow{w} \Sigma F$$

where there are no phantom maps out of F or $\coprod_\alpha X_\alpha$. Now if $g\colon X \to Y$ is a phantom map, then $gv = 0$ so $g = f'w$, say. If $h\colon Y \to Z$ is another phantom map, then hf' is a phantom map out of ΣF, so is trivial. Thus $hg = hf'w$ is trivial as well. □

4.3. **Smashing localizations of Brown categories.** In this section we show that a smashing localization of a Brown category is again a Brown category, and we also prove some slightly sharper statements.

Let $\mathcal{C}$ be a Brown category, and L a smashing localization functor on $\mathcal{C}$. As usual, we write C for the corresponding colocalization functor, and J for the inclusion of $\mathcal{C}_L$ in $\mathcal{C}$. We know by Theorem 3.5.2 that $\mathcal{C}_L$ is an algebraic stable homotopy category, and that $L\colon \mathcal{C} \to \mathcal{C}_L$ is a geometric morphism that preserves small objects.

Proposition 4.3.1. *Let X be an object of $\mathcal{C}_L$. Then X is the minimal weak colimit of the objects LY_α, where Y_α runs over $\Lambda(JX)$. Moreover, X is small in $\mathcal{C}_L$ if and only if it is a retract of LY for some small object Y in $\mathcal{C}$.*

Proof. Suppose that X is small in $\mathcal{C}_L$. By Theorem 4.2.4, we see that JX is the minimal weak colimit of the Y_α. As L preserves minimal weak colimits (by Theorem 3.5.1), we see that $X = LJX$ is the minimal weak colimit of the LY_α.

Now suppose that X is small. As $[X, -]$ is a homology theory on $\mathcal{C}_L$, we see that $[X, X] = \varinjlim_\alpha [X, LY_\alpha]$. This means that the identity map of X factors through some LY_α, so that X is a retract of LY_α, and Y_α is small in $\mathcal{C}$.

Conversely, we know that L preserves smallness (by Theorem 3.5.2), so any retract of a localization of a small object is small. □

We now write $\mathcal{P}_L$ for the ideal of phantoms in $\mathcal{C}_L$. We let $\mathcal{C}/\mathcal{P}$ be the quotient category of $\mathcal{C}$ in which the phantoms are sent to zero, so that $V_\bullet$ induces an equivalence $\mathcal{C}/\mathcal{P} \simeq \mathcal{C}_\bullet$.

Proposition 4.3.2. *Consider objects $X \in \mathcal{C}$ and $U \in \mathcal{C}_L$. Then a map $f\colon X \to JU$ is phantom in $\mathcal{C}$ if and only if the adjoint map $g\colon LX \to U$ is phantom in $\mathcal{C}_L$. Thus, the functors L and J induce an adjoint pair of functors between $\mathcal{C}/\mathcal{P}$ and $\mathcal{C}_L/\mathcal{P}_L$.*

Proof. Suppose that f is phantom. Consider a map $u\colon W \to LX$, where W is small in $\mathcal{C}_L$. We can write X as the minimal weak colimit of a diagram of small objects X_α in $\mathcal{C}$, so LX is the minimal weak colimit of the LX_α, so u factors through some LX_α. However, $LX_\alpha \to LX \xrightarrow{g} U$ is adjoint to $X_\alpha \to X \xrightarrow{f} JU$, which is zero as f is phantom. Thus $gu = 0$ for all such u, which means that g is phantom.

Conversely, suppose that g is phantom. For any small Z in $\mathcal{C}$ and any map $Z \to X$, the composite $Z \to X \xrightarrow{f} JU$ is adjoint to $LZ \to LX \xrightarrow{g} U$, which is zero as LZ is small in $\mathcal{C}_L$. This means that f is phantom.

The second statement of the proposition follows easily. □

Given an object $X \in \mathcal{C}$, we write H_X for the represented homology functor $\mathcal{C} \to \mathrm{Ab}$, defined using small objects in $\mathcal{C}$. Given an object $U \in \mathcal{C}_L$, we write H^L_U for the represented homology functor $\mathcal{C}_L \to \mathrm{Ab}$, defined using small objects in $\mathcal{C}_L$.

Proposition 4.3.3. *For objects $X \in \mathcal{C}$ and $U \in \mathcal{C}_L$, there are natural isomorphisms $H_X \circ J = H^L_{LX}$ and $H^L_U \circ L = H_{JU}$.*

Proof. We first show that $H_X \circ J = H^L_{LX}$. Consider a small object Z in $\mathcal{C}$. Then LZ is small in $\mathcal{C}_L$, so $H^L_{LX} LZ = \pi_0(LX \wedge LZ) = \pi_0(LX \wedge Z)$ (because $LX \wedge LZ = LX \wedge LS \wedge Z = LLX \wedge Z = LX \wedge Z$). On the other hand,

$$H_X JLZ = \widehat{\pi}_0(X \wedge LS \wedge Z) = \widehat{\pi}_0(LX \wedge Z) = \pi_0(LX \wedge Z)$$

(using the smallness of Z). There is thus an isomorphism $H_X JLZ = H^L_{LX} LZ$, natural in Z. Moreover, we can write any $W \in \mathcal{C}_L$ as the minimal weak colimit over $\Lambda(JW)$ of the objects LZ_α, in a functorial way; it follows that there is an isomorphism

$$H_X JW = H^L_{LX} W = \varinjlim_{\Lambda(JW)} \pi_0(LX \wedge Z_\alpha),$$

natural in W. Thus $H_X \circ J = H^L_{LX}$ as claimed.

We now show that $H^L_U \circ L = H_{JU}$. It is enough to check this on small objects of $\mathcal{C}$; let Z be such an object. Then $H_{JU} Z = \pi_0(U \wedge Z)$. Moreover, LZ is small in $\mathcal{C}_L$, so $H^L_U LZ = \pi_0(U \wedge LZ)$. As $U \wedge LZ = U \wedge LS \wedge Z = U \wedge Z$, this is the same as $H_{JU} Z$, as required. □

Theorem 4.3.4. *Let $\mathcal{C}$ be a Brown category and L a smashing localization functor. Then $\mathcal{C}_L$ is a Brown category. Moreover, there is a commutative diagram of functors as follows:*

$$\begin{array}{ccccc}
\mathcal{C}_L/\mathcal{P}_L & \xrightarrow{J} & \mathcal{C}/\mathcal{P} & \xrightarrow{L} & \mathcal{C}_L/\mathcal{P}_L \\
{\scriptstyle V^L_\bullet}\downarrow & & {\scriptstyle V_\bullet}\downarrow & & \downarrow{\scriptstyle V^L_\bullet} \\
\mathcal{C}_{L\bullet} & \xrightarrow[L^*]{} & \mathcal{C}_\bullet & \xrightarrow[J^*]{} & \mathcal{C}_{L\bullet}
\end{array}$$

Here $V^L_\bullet$ is the functor $U \mapsto H^L_U$, L^* is the functor $H \mapsto H \circ L$, and everything else should be clear. The vertical functors are equivalences, the functors L^* and J are full and faithful, and the functors L and J^* are essentially surjective. The horizontal composites are identity functors.

Proof. Proposition 4.3.2 tells us that the functors marked J and L are well-defined. Proposition 4.3.3 says that the squares commute up to natural isomorphism. As $LJ = 1$, the horizontal composites are identity functors. It follows that L^* and J are full and faithful, and that L and J^* are essentially surjective. As $\mathcal{C}$ is a Brown category, the functor $V_\bullet$ is an equivalence. As three sides of the left hand square are full and faithful, it follows that the fourth side $V_\bullet^L$ is full and faithful. As $V_\bullet^L \circ L = J^* \circ V_\bullet$ is essentially surjective, the same is true of $V_\bullet^L$. Thus $V_\bullet^L$ is an equivalence, as claimed. □

4.4. **A topology on** $[X, Y]$**.** In this brief section, we point out that there is a natural topology on the morphisms in an algebraic stable homotopy category $\mathcal{C}$, which enriches the category over topological Abelian groups.

Recall that a *linear topology* on an Abelian group A is a topology such that the cosets of open subgroups form a basis of open sets. Given a family $\{A_i\}$ of subgroups of A, there is a unique linear topology on A such that subgroups A_i are open and form a basis of neighborhoods of 0. We write $A' \leq_O A$ to indicate that A' is an open subgroup of A.

There is a natural map $\alpha\colon A \longrightarrow \varprojlim_{A' \leq_O A} A/A'$. We shall say that A is complete if α is surjective. Moreover, A is Hausdorff if and only if α is injective, if and only if the intersection of the open subgroups is zero, if and only if $\{0\}$ is closed.

Fix two objects X and Y of $\mathcal{C}$. For any map $F \xrightarrow{f} X$ from a small object to X, let $U_f = U_f(X, Y)$ denote the kernel of $f^*\colon [X, Y] \longrightarrow [F, Y]$. We give $[X, Y]$ the linear topology determined by the subgroups U_f, and refer to this as the *natural topology*.

Proposition 4.4.1. *Let $\mathcal{C}$ be an algebraic stable homotopy category.*

(a) *The composition map $[X, Y] \times [Y, Z] \longrightarrow [X, Z]$ is continuous.*
(b) *Any pair of maps $X' \longrightarrow X$ and $Y \longrightarrow Y'$ induces a continuous map $[X, Y] \longrightarrow [X', Y']$.*
(c) *If X is small then $[X, Y]$ is discrete.*
(d) *The closure of 0 in $[X, Y]$ is the set of phantom maps, so $[X, Y]$ is Hausdorff if and only if $\mathcal{P}(X, Y) = 0$.*
(e) *If $\mathcal{C}$ is a Brown category then $[X, Y]$ is always complete.*
(f) *$[\coprod_i X_i, Y]$ is homeomorphic to $\prod_i [X_i, Y]$ with the product topology, but the natural topology on $[X, \prod Y_i]$ is strictly finer than the product topology in general.*

Proof. (a): Call the composition map γ. Suppose that we have maps $X \xrightarrow{u} Y \xrightarrow{v} Z$, and a neighborhood $vu + U_f$ of $\gamma(u, v)$ in $[X, Z]$ (so $f\colon F \longrightarrow X$ for some small F). Then one sees easily that $\gamma((u + U_f) \times (v + U_{uf})) \subseteq vu + U_f$, so that γ is continuous at (u, v).

(b): This follows immediately from (*a*).

(c): Immediate from the definitions.

(d): Immediate from the definitions.

(e): This is equivalent to the statement that X is a weak colimit of $\Lambda(X)$, which is Theorem 4.2.4.

(f): The first part is easy to see using the fact that any map from a small object to a coproduct factors through a finite sub-coproduct. For the second part, suppose that we have a map $f\colon F \to X$ from a small object F to X. Then

$$U_f(X, \prod_i Y_i) = \prod_i U_f(X, Y_i).$$

This product is therefore open in the natural topology, but rarely in the product topology. □

This construction gives an enrichment of $\mathcal{C}$ over topological Abelian groups. This becomes very important in the $K(n)$-local category, where even homotopy groups can have interesting topology [HSS]. If S is small then many of the groups which arise are discrete.

We should point out that Brown representability is not compatible with this enrichment. That is, there are cohomology functors to the category of topological Abelian groups which are not representable. Indeed, given an infinite family $\{Y_i\}$ of objects in a stable homotopy category $\mathcal{C}$, define a cohomology functor H by $H(X) = \prod[X, Y_i]$ with the product topology. This functor cannot be representable, since if it were, $H(X)$ would have to have the discrete topology for all small X. In general, a cohomology functor H is representable if and only if, for all X, a subset U of $H(X)$ is a neighborhood of 0 if and only if there exists a small object F and a map $f\colon F \to X$ such that $U \supseteq H(f)^{-1}(0)$.

5. Nilpotence and thick subcategories

In this section we present analogues of the nilpotence theorems of Devinatz, Hopkins and Smith [DHS88], and the thick subcategory theorems of Hopkins and Smith [HS]. None of our theorems imply the theorems just mentioned; we require stronger finiteness conditions than are available in their context. Our nilpotence theorems require a unital algebraic stable homotopy category $\mathcal{C}$, but our thick subcategory theorems will hold in an arbitrary algebraic stable homotopy category. We allow $\mathcal{C}$ to be multigraded, and we will often consider graded maps—see Section 1.3.

5.1. A naïve nilpotence theorem.

In this section we prove a nilpotence theorem. We first present some miscellaneous definitions.

Definition 5.1.1.

(a) We write $X^{(m)} = X \wedge \cdots \wedge X$ (with m factors), and similarly for maps.
(b) We say that a graded map $f\colon X \to Y$ is *smash nilpotent* if $f^{(m)}\colon X^{(m)} \to Y^{(m)}$ is null for $m \gg 0$.
(c) Suppose that we have a map $f\colon S \to \Sigma^{-d}X$. We then get a sequence

$$S = X^{(0)} \xrightarrow{f} \Sigma^{-d}X^{(1)} \xrightarrow{f\wedge 1} \Sigma^{-2d}X^{(2)} \dots . \tag{5.1.1}$$

We write $X^{(\infty)}$ for the sequential colimit, and $f^{(\infty)}$ for the evident map $S \to X^{(\infty)}$.
(d) We say that a graded self-map $f\colon X \to X$ is *composition nilpotent* or just *nilpotent* if the mth composition power $f^m\colon X \to X$ is null for $m \gg 0$.

Here is a rather generic nilpotence theorem. See Definitions 2.1.1 and 3.6.1 for the relevant definitions.

Theorem 5.1.2 (Nilpotence theorem I). *Let $\mathcal{C}$ be a unital algebraic stable homotopy category. Suppose that we have objects $\{K(n) \mid n \in I\}$ (for some indexing set I), so that*

$$(5.1.2) \qquad \langle S \rangle = \coprod_{n \in I} \langle K(n) \rangle.$$

Then the objects $K(n)$ detect nilpotence:

(a) *Let F be small, and X arbitrary. A graded map $f\colon F \longrightarrow X$ is smash nilpotent if $1_{K(n)} \wedge f = 0$ for all n.*
(b) *Let X be a small object. A graded map $f\colon X \longrightarrow X$ is nilpotent if $1_{K(n)} \wedge f = 0$ for all n.*
(c) *Suppose that each $K(n)$ is monoidal, and let X be a small object. A graded map $f\colon X \longrightarrow X$ is nilpotent if and only if for each n, the map $K(n)_*(f)$ is nilpotent.*

Equation (5.1.2) simply means that whenever $K(n) \wedge X = 0$ for all n, we have $X = 0$. For part (c), we can always replace $K(n)$ by a Bousfield-equivalent monoidal object, using Lemma 3.6.8. If the objects $K(n)$ are ring objects (Definition 3.7.1), then we have results more like those in [HS].

Theorem 5.1.3 (Nilpotence theorem II). *Suppose, in addition, that each $K(n)$ is a monoidal ring object.*

(a) *Let R be a ring object. An element $\alpha \in \pi_* R$ is nilpotent if and only if $K(n)_*(\alpha)$ is nilpotent for all n.*
(b) *Let X be a small object. A graded map $f\colon S \longrightarrow X$ is smash nilpotent if $K(n)_*(f) = 0$ for all n.*

Remark 5.1.4.

(a) Neither of these results implies the nilpotence theorems of [DHS88] and [HS], because Equation (5.1.2) does not hold in the homotopy category of spectra, with MU, BP, or the wedge of the Morava K-theories on the right hand side. Theorems 5.1.2 and 5.1.3 seem to be useful mainly in a stable homotopy category satisfying certain strong finiteness conditions.
(b) Note that Theorem 5.1.3(b) is weaker than the smash nilpotence result in [HS]. Hopkins and Smith's theorem applies to maps $f\colon F \longrightarrow X$ where F is small. They reduce to the case $F = S$ using Spanier-Whitehead duality to convert f to a map $\hat{f}\colon S \longrightarrow DF \wedge X$, and the Künneth isomorphism for $K(n)_*$ to see that $K(n)_*(f) = 0 \Rightarrow K(n)_*(\hat{f}) = 0$. This application of the Künneth isomorphism seems to be necessary, and one cannot expect it to hold in an arbitrary stable homotopy category.

We need three lemmas before we begin the proofs.

Lemma 5.1.5. (a) *A graded map $f\colon S \longrightarrow X$ is smash nilpotent if and only if $X^{(\infty)}$ is contractible.*
(b) *If $f\colon S \longrightarrow X$ and E are such that $1_E \wedge f = 0$, then $E \wedge X^{(\infty)} = 0$.*

Proof. (a): Certainly if f is smash nilpotent, then $X^{(\infty)}$ is contractible. Conversely, we have

$$\begin{aligned} X^{(\infty)} \simeq 0 &\Rightarrow [S, X^{(\infty)}]_* = 0 \\ &\Rightarrow S \to X \to X^{(2)} \to \dots \to X^{(\infty)} \text{ is null} \\ &\Rightarrow S \to X \to X \wedge X \to \dots \to X^{(n)} \text{ is null for } n \gg 0 \\ &\Rightarrow f \text{ is smash nilpotent.} \end{aligned}$$

(b): Smash the diagram (5.1.1) with E. The sequential colimit is $E \wedge X^{(\infty)}$, and each map in the diagram is null. □

Lemma 5.1.6. *Suppose that $f\colon \Sigma^d X \to X$ is a graded self-map of a small object X. Recall (Definition 2.2.3) that $f^{-1}X$ denotes the sequential colimit of the sequence $X \xrightarrow{f} \Sigma^{-d}X \to \Sigma^{-2d}X \to \dots$. Let $E \in \mathcal{C}$ be any object.*

(a) *f is nilpotent if and only if $f^{-1}X = 0$.*
(b) *If $E_*(f)$ is nilpotent, then $E_*(f^{-1}X) = 0$.*

Proof. (a): Certainly if f is nilpotent, then $f^{-1}X$ is trivial. On the other hand, since X is small we have $[X, f^{-1}X] = \varinjlim [X, X]$, where the maps in the colimit are composition with f. Hence if $f^{-1}X = 0$, the identity map of X must be 0 at a finite stage of the colimit. In other words, f must be nilpotent.

(b): If $E_*(f)$ is nilpotent, say $E_*(f^r) = 0$, then the maps in the sequence

$$E \wedge X \xrightarrow{1 \wedge f^r} E \wedge X \xrightarrow{1 \wedge f^r} \dots$$

all induce zero on the homology functor $[S, -]_*$. Hence the minimal weak colimit is S_*-acyclic. In other words, $E_*(f^{-1}X) = 0$. □

Proof of Theorem 5.1.2. For part (a), we use Lemma 5.1.5. First, using Spanier-Whitehead duality, we can reduce to the case where $F = S$, so assume we have a map $f\colon S \to X$. Let $X^{(\infty)}$ be as above. By assumption, $1_{K(n)} \wedge f = 0$ for all n, so $K(n) \wedge X^{(\infty)} = 0$ for all n. As $\langle S \rangle = \coprod_n \langle K(n) \rangle$, we conclude that $X^{(\infty)} = 0$. Thus f is smash nilpotent.

Part (b) follows from (a), as in [HS].

For part (c), we use Lemma 5.1.6, so we have to show that $f^{-1}X = 0$. The hypothesis implies immediately that $K(n)_* f^{-1}X = 0$; as $K(n)$ is monoidal, we see that $K(n) \wedge f^{-1}X = 0$. As this holds for all n, we have $f^{-1}X = 0$. □

Suppose that R is a ring object. Fix $\alpha \in \pi_*(R)$, and let $\hat{\alpha}$ denote the "multiplication by α" self-map $\mu \circ (\alpha \wedge 1)$ of R.

Lemma 5.1.7. *With notation as above, α is nilpotent in $\pi_*(R)$ if and only if $\hat{\alpha}^{-1}R = 0$.*

Proof. Because $\mathcal{C}$ is unital algebraic, we know that $\hat{\alpha}^{-1}R = 0$ if and only if

$$\pi_*(\hat{\alpha}^{-1}R \wedge DZ) = 0$$

for all $Z \in \mathcal{G}$. The right hand side is just the direct limit of the R-module $\pi_*(R \wedge DZ)$ under left multiplication by α. This vanishes if α is nilpotent, and the converse holds by the special case $Z = S$. □

Proof of Theorem 5.1.3. Part (a) follows from Lemma 5.1.7—if $\beta = K(n)_*(\alpha)$ is nilpotent, then

$$K(n) \wedge \hat{\alpha}^{-1}R = \hat{\beta}^{-1}(K(n) \wedge R) = 0.$$

If this holds for all n, then by the decomposition (5.1.2), we see that $\hat{\alpha}^{-1}R = 0$, so α is nilpotent.

(b): Let $\eta\colon S \longrightarrow K(n)$ denote the unit map, and $\mu\colon K(n) \wedge K(n) \longrightarrow K(n)$ the product. Suppose that $f\colon S \longrightarrow X$ induces zero on $K(n)_*$; then the composite $S \xrightarrow{\eta} K(n) \xrightarrow{1\wedge f} K(n) \wedge X$ is null. But $K(n) \xrightarrow{1\wedge f} K(n) \wedge X$ factors as $K(n) \wedge S \xrightarrow{(1\wedge f)\circ\eta} K(n) \wedge K(n) \wedge X \xrightarrow{\mu\wedge 1} K(n) \wedge X$, and so is null. Now apply Theorem 5.1.2(b). □

5.2. **A thick subcategory theorem.** In this section we present a classification of the $\mathcal{G}$-ideals of small objects in an algebraic stable homotopy category; the basic argument is of course inspired by [HS], and a few of the details are drawn from [Ric].

Definition 5.2.1. Suppose that $\mathcal{C}$ is a stable homotopy category. Fix a collection of objects $\{K(n) \mid n \in I\}$ (for some indexing set I). Given an object X, we define the *support* of X (with respect to the $K(n)$'s) to be the set $\operatorname{supp}(X) = \{n \mid K(n) \wedge X \neq 0\}$. Similarly, given a replete subcategory $\mathcal{D}$, define the support of $\mathcal{D}$ to be the set $\operatorname{supp}(\mathcal{D}) = \bigcup_{X\in\mathcal{D}} \operatorname{supp} X$. We say that the $K(n)$'s *determine* $\mathcal{G}$*-ideals* if whenever $\mathcal{D}$ is a $\mathcal{G}$-ideal of small objects, we have

$$\mathcal{D} = \{X \mid X \text{ finite}, \operatorname{supp}(X) \subseteq \operatorname{supp}(\mathcal{D})\}.$$

Theorem 5.2.2. *Suppose that $\mathcal{C}$ is an algebraic stable homotopy category, and that we have objects $\{K(n)\}$ such that*

1. *If R is a nontrivial ring object, then there is some n such that $K(n) \wedge R$ is nontrivial (in other words, the objects $K(n)$ detect ring objects).*
2. *If X is finite and $K(n) \wedge X \neq 0$, then $\langle K(n)\rangle = \langle K(n) \wedge X\rangle$.*

Then the objects $K(n)$ determine $\mathcal{G}$-ideals.

Note that we always have $\langle K(n)\rangle \geq \langle K(n) \wedge X\rangle$. If $\mathcal{C}$ is monogenic, then every thick subcategory is a $\mathcal{G}$-ideal, and vice versa, so we get a classification of thick subcategories in this setting.

Note as well that any collection of ring objects that detect nilpotence, as in Theorem 5.1.3, automatically detects ring objects. Theorem 5.2.2 tells us what else we need to know to get a thick subcategory theorem from a nilpotence theorem. In particular, we recover the Hopkins-Smith thick subcategory theorem [HS] from the Devinatz-Hopkins-Smith nilpotence theorem [DHS88].

Corollary 5.2.3. *Suppose that $\mathcal{C}$ is monogenic, so that thick subcategories are the same as $\mathcal{G}$-ideals. If the family $\{K(n)\}$ detects ring objects and each $K(n)$ satisfies one of the following conditions, then the $K(n)$'s determine thick subcategories:*

(i) *For X and Y arbitrary objects, $K(n)_*(X\wedge Y) = 0$ if and only if $K(n)_*(X) = 0$ or $K(n)_*(Y) = 0$.*
(ii) *$K(n)_*$ satisfies a Künneth isomorphism: $K(n)_*(X \wedge Y) \simeq K(n)_*(X) \otimes_{K(n)_*} K(n)_*(Y)$.*
(iii) *$K(n)$ is a skew field object (Definition 3.7.1).*
(iv) *$\langle K(n)\rangle$ is a minimal nonzero Bousfield class.* □

Proof of Theorem 5.2.2. Suppose that $\mathcal{D}$ is an $\mathcal{G}$-ideal, and $\operatorname{supp}(Y) \subseteq \operatorname{supp}(\mathcal{D})$. We need to show that $Y \in \mathcal{D}$. Note first that Y and $F(Y,Y) = Y \wedge DY$ generate the same thick subcategory (by Lemma A.2.6), so we can replace Y by the ring object $F(Y,Y)$. We therefore assume that Y is a ring object.

Since $\mathcal{C}$ is algebraic, every $\mathcal{G}$-ideal of small objects is essentially small, so we can use the finite localization functors of Definition 3.3.4. Thus, it suffices to show that $Y \wedge L^f_{\mathcal{D}} S = L^f_{\mathcal{D}} Y = 0$. Fix n. If $K(n) \wedge Y = 0$, then certainly $K(n) \wedge Y \wedge L^f_{\mathcal{D}} S = 0$. If $K(n) \wedge Y \neq 0$, then (because $\operatorname{supp}(Y) \subseteq \operatorname{supp}(\mathcal{D})$) there is some $X \in \mathcal{D}$ such that $K(n) \wedge X \neq 0$. Since $X \wedge L^f_{\mathcal{D}} S = 0$, we have $(K(n) \wedge X) \wedge (Y \wedge L^f_{\mathcal{D}} S) = 0$; and since $\langle K(n) \rangle = \langle K(n) \wedge X \rangle$, we have $K(n) \wedge (Y \wedge L^f_{\mathcal{D}} S) = 0$. So for all n, we have $K(n) \wedge (Y \wedge L^f_{\mathcal{D}} S) = 0$; hence, since $Y \wedge L^f_{\mathcal{D}} S$ is a ring object, it is trivial. □

6. Noetherian stable homotopy categories

In this section we consider a multigraded stable homotopy category $\mathcal{C}$—see Section 1.3. We shall assume that $\mathcal{C}$ is monogenic in the multigraded sense.

Definition 6.0.1. If $\mathcal{C}$ is a monogenic stable homotopy category such that $\pi_*(S) = [S,S]_*$ is Noetherian (as a multigraded-commutative ring), then we say that $\mathcal{C}$ is a *Noetherian* stable homotopy category.

We shall use a number of standard theorems that are proved in the literature for ungraded commutative rings. These will all apply to multigraded-commutative rings, when suitably interpreted. On the one hand, one has to keep track of the grading, but this is trivial if we insist that everything in sight be homogeneous. On the other hand, we need to think about the fact that odd-dimensional elements anticommute instead of commuting. If 2 is invertible in R, then all odd-dimensional elements square to zero and thus lie in every prime ideal, and this makes everything work as expected. If we work modulo 2, then R is strictly commutative. By combining these observations, we see that everything works as expected integrally. The very cautious reader may wish to assume that R is concentrated in even degrees, as we could not honestly claim to have checked every detail otherwise.

The derived category of a Noetherian ring is a Noetherian stable homotopy category. If B is a finite-dimensional commutative Hopf algebra, then $\mathcal{C}(B)$ is often a Noetherian stable homotopy category (see Section 9.5). In particular, this holds if $B = (kG)^*$ with G a finite p-group and $\operatorname{char}(k) = p$, or if B is graded and connected.

For the rest of this section, we assume that $\mathcal{C}$ is a Noetherian stable homotopy category. We write R for the graded ring $\pi_* S = [S,S]_*$. We also let $\operatorname{Spec} R$ (respectively, $\operatorname{Max} R$) denote the space of prime (maximal) homogeneous ideals of R, under the Zariski topology. Given an R-module M and a prime ideal $\mathfrak{p} \in \operatorname{Spec} R$ we write the localization, completion and completed localization of M at $\mathfrak{p}$ as follows:

$$M_{\mathfrak{p}} = (R \setminus \mathfrak{p})^{-1} M$$
$$M^{\wedge}_{\mathfrak{p}} = \varprojlim{}_k M/\mathfrak{p}^k M$$
$$M^{\diamond}_{\mathfrak{p}} = (M_{\mathfrak{p}})^{\wedge}_{\mathfrak{p}}$$

(We need $M^{\diamond}_{\mathfrak{p}}$ because $M^{\wedge}_{\mathfrak{p}}$ need not be $\mathfrak{p}$-local in general. There does not seem to be a standard notation for this, so we have invented one.)

We also write $E(R/\mathfrak{p})$ for the injective hull of $R/\mathfrak{p}$, which is well-defined up to non-canonical isomorphism.

We record some basic facts from commutative algebra.

Proposition 6.0.2. *$R^{\diamond}_{\mathfrak{p}}$ and $R_{\mathfrak{p}}$ are flat over R, and $R^{\diamond}_{\mathfrak{p}}$ is faithfully flat over $R_{\mathfrak{p}}$. For any M, we have $M_{\mathfrak{p}} = R_{\mathfrak{p}} \otimes_R M$; hence $M \mapsto M_{\mathfrak{p}}$ is an exact functor. If M is finitely generated, we have $M^{\wedge}_{\mathfrak{p}} = R^{\wedge}_{\mathfrak{p}} \otimes_R M$ and $M^{\diamond}_{\mathfrak{p}} = R^{\diamond}_{\mathfrak{p}} \otimes_R M$.* □

Proposition 6.0.3. *A module M is zero if and only if $M_{\mathfrak{m}} = 0$ for all $\mathfrak{m} \in \operatorname{Max}(R)$, if and only if $M_{\mathfrak{p}} = 0$ for all $\mathfrak{p} \in \operatorname{Spec} R$.* □

The following less well-known result is proved in [Mat89, Section 18].

Proposition 6.0.4. *$E(R/\mathfrak{p})$ is a $\mathfrak{p}$-torsion module; more precisely, for all $x \in E(R/\mathfrak{p})$, there is a $t \geq 1$ so that $\mathfrak{p}^t x = 0$. The localizations of $E(R/\mathfrak{p})$ are*

$$E(R/\mathfrak{p})_{\mathfrak{q}} = \begin{cases} E(R/\mathfrak{p}) & \mathfrak{p} \leq \mathfrak{q}, \\ 0 & \text{otherwise.} \end{cases}$$

Moreover, there is a natural isomorphism $R^{\diamond}_{\mathfrak{p}} = \operatorname{Hom}_R(E(R/\mathfrak{p}), E(R/\mathfrak{p}))$ □

The following fact is standard, but seldom stated explicitly.

Proposition 6.0.5. *In a Noetherian ring, any nonempty collection of prime ideals has a minimal element.*

Proof. If R is a Noetherian local ring, then it has finite Krull dimension (bounded by $\dim_{R/\mathfrak{m}} \mathfrak{m}/\mathfrak{m}^2$). Given an arbitrary Noetherian ring R, and a collection T of prime ideals, choose $\mathfrak{p} \in T$ such that the Krull dimension of $R_{\mathfrak{p}}$ (also called the height of $\mathfrak{p}$) is minimal. It is easy to see that any $\mathfrak{q} < \mathfrak{p}$ has strictly smaller height, so $\mathfrak{p}$ is minimal in T. □

We now start to apply these results to stable homotopy theory.

Proposition 6.0.6. *If X and Y are small objects in a Noetherian stable homotopy category $\mathcal{C}$, then $[X,Y]_*$ is finitely generated as a module over $R = \pi_* S$.*

Proof. By Spanier-Whitehead duality it suffices to show that if Y is small, then $\pi_* Y$ is a finitely generated R-module. This is proved by showing that the category of all Y such that $\pi_* Y$ is a finitely generated R-module is thick, which follows from well-known properties of finitely-generated modules over a Noetherian ring. □

For example, if X is small, then the noncommutative ring $[X,X]_*$ is finitely generated as a module over the image of $R \to [X,X]_*$.

A basic technique in Noetherian ring theory is to work one prime at a time by localizing. Fortunately, we have an analogous procedure in Noetherian stable homotopy theory.

Proposition 6.0.7. *For each $\mathfrak{p} \in \operatorname{Spec} R$, there is a ring object $S_{\mathfrak{p}}$ such that $\pi_*(S_{\mathfrak{p}} \wedge X) = \pi_*(X)_{\mathfrak{p}}$. Moreover, the functor $L_{\mathfrak{p}} \colon X \mapsto X_{\mathfrak{p}} = S_{\mathfrak{p}} \wedge X$ is an algebraic localization.*

Proof. This is an immediate consequence of Theorem 3.3.7 and Proposition 3.1.8. □

It follows that the category of local objects, which we denote $\mathcal{C}_{\mathfrak{p}}$, is again a Noetherian stable homotopy category. We call it the $\mathfrak{p}$-*localization* of $\mathcal{C}$.

Now we define a number of other objects associated to a prime ideal $\mathfrak{p} \leq R = \pi_* S$.

Definition 6.0.8. Fix a prime ideal $\mathfrak{p} \leq R$.

(a) $S/\mathfrak{p}$: Write the prime ideal $\mathfrak{p}$ as $\mathfrak{p} = (y_1, \dots, y_n)$. Define S/y_i as the cofiber of the (graded) map $S \xrightarrow{y_i} S$, and let
$$S/\mathfrak{p} = S/y_1 \wedge \cdots \wedge S/y_n.$$
Note that $S/\mathfrak{p}$ depends on the choice of generators $\{y_i\}$. We shall show in Lemma 6.0.9 that any two choices generate the same thick subcategory, and thus have the same Bousfield class.

(b) $K(\mathfrak{p})$: Let $K(\mathfrak{p}) = S_\mathfrak{p} \wedge S/\mathfrak{p} = (S/\mathfrak{p})_\mathfrak{p}$. Using Lemma A.2.6, we see that this generates the same thick subcategory as the ring object $F(K(\mathfrak{p}), K(\mathfrak{p})) = S_\mathfrak{p} \wedge D(S/\mathfrak{p}) \wedge S/\mathfrak{p}$.

(c) $I_\mathfrak{p}$: As $E(R/\mathfrak{p})$ is an injective module, the functor
$$X \mapsto \operatorname{Hom}_R(\pi_* X, E(R/\mathfrak{p}))$$
is a cohomology functor. We let $I_\mathfrak{p}$ denote the representing object. Note that $\pi_* I_\mathfrak{p} = E(R/\mathfrak{p})$.

(d) $S_\mathfrak{p}^\diamond$: Define $S_\mathfrak{p}^\diamond = F(I_\mathfrak{p}, I_\mathfrak{p})$. Note that $\pi_* S_\mathfrak{p}^\diamond = R_\mathfrak{p}^\diamond$, by Proposition 6.0.4.

(e) $M_\mathfrak{p} S$: Let $L_\mathfrak{p}$ be the $\mathfrak{p}$-localization functor; this is a finite localization, and the corresponding subcategory of finite acyclics is
$$\{Z \in \mathcal{F} \mid \pi_*(Z)_\mathfrak{p} = 0\} = \{Z \in \mathcal{F} \mid \pi_*(Z)_\mathfrak{q} = 0 \text{ for all } \mathfrak{q} \leq \mathfrak{p}\}.$$
We let $L_{<\mathfrak{p}}$ be the finite localization determined by the set of small objects Z with $\pi_*(Z)_\mathfrak{q} = 0$ for $\mathfrak{q} < \mathfrak{p}$. There is then a unique morphism $L_\mathfrak{p} \longrightarrow L_{<\mathfrak{p}}$, and we write $M_\mathfrak{p}$ for the fiber. The objects $M_\mathfrak{p} X$ are used in [BCR].

(See also Proposition 9.3.2 for yet another object associated to $\mathfrak{p}$. This one is only defined when working in the derived category of a ring; it is a field, and its coefficient ring is the residue field of $\mathfrak{p}$.)

We owe the reader a proof that $S/\mathfrak{p}$ is essentially well-defined.

Lemma 6.0.9. *The thick subcategory generated by $S/\mathfrak{p}$ (and thus its Bousfield class) is independent of the choices made. Moreover, if $\mathfrak{p} \leq \mathfrak{q}$ then $S/\mathfrak{q} \in \operatorname{thick}\langle S/\mathfrak{p}\rangle$.*

Proof. Suppose that we chose generators $(y_1, \dots, y_n)$ for $\mathfrak{p}$, so the resulting model of $S/\mathfrak{p}$ is $Y = \bigwedge_i S/y_i$. One can check that $y_i^2 = 0$ as a self-map of $S/(y_i)$, and thus that every element of $\mathfrak{p}$ acts nilpotently on Y. Now choose a different set of generators $(z_1, \dots, z_m)$. For large N we see that each z_i^N acts trivially on Y, and thus that Y is a retract of $Y \wedge S/(z_1^N, \dots, z_m^N)$. It follows easily that Y lies in the thick subcategory generated by $Z = S/(z_1, \dots, z_m)$, as required. An evident extension of this argument gives the second claim. □

6.1. **Monochromatic subcategories.** Let $\mathcal{C}$ be a Noetherian stable homotopy category. We will need to investigate certain subcategories that are strongly concentrated at a single prime ideal $\mathfrak{p}$ of $R = \pi_*(S)$.

Definition 6.1.1. The *monochromatic category* $\mathcal{M}_\mathfrak{p}$ is the localizing subcategory generated by $K(\mathfrak{p})$.

Theorem 6.1.8 gives a number of other descriptions of this category.

We make the following conjectures:

Conjecture 6.1.2. *For each prime $\mathfrak{p}$, the category $\mathcal{M}_\mathfrak{p}$ is minimal among nonzero localizing subcategories of $\mathcal{C}$.*

Conjecture 6.1.3. *For each $\mathfrak{p}$, the Bousfield class $\langle K(\mathfrak{p})\rangle$ is minimal among non-trivial Bousfield classes.*

We shall show in Proposition 6.1.7 and Theorem 6.1.9 that

$$\langle S\rangle = \amalg_{\mathfrak{p}}\langle K(\mathfrak{p})\rangle, \quad \text{and}$$
$$K(\mathfrak{p}) \wedge K(\mathfrak{q}) = 0 \quad \text{for} \quad \mathfrak{p} \neq \mathfrak{q}.$$

This means that each $K(\mathfrak{p})$ is smash-complemented. Thus, if each $K(\mathfrak{p})$ is (or generates the same thick subcategory as) a skew field object, then both conjectures follow from Proposition 3.7.3. We shall show in Proposition 6.1.11 that Conjecture 6.1.2 implies Conjecture 6.1.3.

Example 6.1.4. Both conjectures hold in the derived category $\mathcal{D}(R)$ of a Noetherian ring R. Indeed, let $\overline{k_{\mathfrak{p}}}$ denote (the representative in $\mathcal{D}(R)$ of) the residue field $k_{\mathfrak{p}} = (R/\mathfrak{p})_{\mathfrak{p}}$ at $\mathfrak{p}$. We will see in Proposition 9.3.2 that $\operatorname{loc}\langle K(\mathfrak{p})\rangle = \operatorname{loc}\langle \overline{k_{\mathfrak{p}}}\rangle$ and $\langle K(\mathfrak{p})\rangle = \langle \overline{k_{\mathfrak{p}}}\rangle$. We will show also that $\overline{k_{\mathfrak{p}}}$ is a smash-complemented field object, so both conjectures follow from Proposition 3.7.3. Hence we recover the thick subcategory theorem of [Hop87, Nee92a] from Theorem 5.2.2. The telescope conjecture also holds in this setting. This was proved in [Nee92a] for the derived category, and will be proved in a more general context in Theorem 6.3.7.

The thick subcategory theorem is also known to be true for $\mathcal{C}((kG)^*)$ when G is a p-group, as proved in [BCR]. Their method is to verify that the objects $M_{\mathfrak{p}}S$ (defined in Definition 6.0.8) satisfy

$$M_{\mathfrak{p}}S \wedge (X \wedge Y) = 0 \Leftrightarrow M_{\mathfrak{p}}S \wedge X = 0 \text{ or } M_{\mathfrak{p}}S \wedge Y = 0.$$

In this context, this also implies that $\langle M_{\mathfrak{p}}\rangle$ is minimal. (They write $\kappa(\mathfrak{p})$ for $M_{\mathfrak{p}}S$, essentially, and [BCR, Theorem 10.8], is precisely the above statement. See also [BCR, Lemma 10.3].) We shall show in Theorem 6.1.8 that $\langle K(\mathfrak{p})\rangle = \langle M_{\mathfrak{p}}S\rangle$.

Assuming Conjecture 6.1.2, we can state our results quite simply. Recall that a subset $T \subseteq \operatorname{Spec} R$ is said to be *closed under specialization* if whenever $\mathfrak{p} \in T$ and $\mathfrak{p} \leq \mathfrak{q}$, we also have $\mathfrak{q} \in T$. This is (easily) equivalent to T being a union of Zariski closed sets. The following theorem is a summary of Theorem 6.2.3, Corollary 6.1.10, Corollary 6.3.4 and Theorem 6.3.7.

Theorem 6.1.5. *Suppose that Conjecture* 6.1.2 *holds for* $\mathcal{C}$. *Then the Bousfield lattice is isomorphic to the lattice of subsets of* $\operatorname{Spec} R$. *Moreover, every generalized Bousfield class is a Bousfield class, and* $\langle X\rangle \cap \langle Y\rangle = \langle X \wedge Y\rangle$ *for all* X *and* Y. *Every smashing localization is a finite localization, and the lattice of such is isomorphic to the lattice of thick subcategories of small objects, or to the lattice of subsets of* $\operatorname{Spec} R$ *that are closed under specialization. The objects* $K(\mathfrak{p})$ *detect nilpotence and determine thick subcategories.* □

We still have strong results in the absence of Conjecture 6.1.2, but they cannot be stated so succinctly. A major rôle is played by the following definition.

Definition 6.1.6. For any object X, we define

$$\operatorname{supp}(X) = \{\mathfrak{p} \mid K(\mathfrak{p})_*(X) \neq 0\} \subseteq \operatorname{Spec} R.$$

Similarly, if $\mathcal{D}$ is a thick subcategory of $\mathcal{C}$, we define

$$\operatorname{supp}(\mathcal{D}) = \bigcup_{X \in \mathcal{D}} \operatorname{supp}(X).$$

For a general object X, this can be an arbitrary subset of $\operatorname{Spec} R$, but for small objects it is constrained by the following result.

Proposition 6.1.7.

(a) *If* $\mathfrak{p} \neq \mathfrak{q}$, *then* $K(\mathfrak{p}) \wedge K(\mathfrak{q}) = 0$.
(b) *Fix a small object* X *and a prime ideal* $\mathfrak{q}$. *Then* $K(\mathfrak{q})_* X = 0$ *if and only if* $X_\mathfrak{q} = 0$. *In particular (taking* $X = S$*), the* $K(\mathfrak{q})$ *are all nontrivial.*
(c) *If* X *is small then* $\operatorname{supp}(X)$ *is Zariski closed (and thus closed under specialization).*
(d) $\operatorname{supp}(S/\mathfrak{p}) = V(\mathfrak{p}) = \{\mathfrak{q} \mid \mathfrak{p} \leq \mathfrak{q}\}$.

Proof. (a): Without loss of generality $\mathfrak{p} \not\leq \mathfrak{q}$, so there exists $y \in \mathfrak{p} \setminus \mathfrak{q}$. As $y \in \mathfrak{p}$, it is nilpotent as a self-map of $K(\mathfrak{p})$, while it is an equivalence on $K(\mathfrak{q})$. It is both nilpotent and an equivalence on $K(\mathfrak{p}) \wedge K(\mathfrak{q})$, so this object must be zero.

(b): Certainly, if $X_\mathfrak{q} = 0$ then $K(\mathfrak{q}) \wedge X = S/\mathfrak{q} \wedge X_\mathfrak{q} = 0$. Conversely, suppose that $S/\mathfrak{q} \wedge X_\mathfrak{q} = 0$. Choose a set of generators $(y_1, \ldots, y_n)$ for $\mathfrak{q}$. We will show by downward induction on i that $Y_i = S/(y_1, \ldots, y_i) \wedge X_\mathfrak{q}$ is trivial. Suppose that Y_k is trivial. By considering the cofibration sequence

$$Y_{k-1} \xrightarrow{y_k} Y_{k-1} \longrightarrow Y_k,$$

we find that multiplication by y_k on $\pi_* Y_{k-1}$ is an isomorphism. Now Y_{k-1} is the $\mathfrak{q}$-localization of a small object. Hence $\pi_* Y_{k-1}$ is a finitely generated module over the Noetherian local ring $R_\mathfrak{q}$ by Proposition 6.0.6. The element y_k is in the Jacobson radical, and $y_k \pi_* Y_{k-1} = \pi_* Y_{k-1}$. Hence $\pi_* Y_{k-1} = 0$ by Nakayama's lemma.

(c): By (b), we have $\operatorname{supp}(X) = \{\mathfrak{p} \mid \pi_*(X)_\mathfrak{p} \neq 0\}$. As $\pi_*(X)$ is finitely generated, a well-known algebraic lemma identifies this with the Zariski closed set $V(\operatorname{ann}(\pi_* X))$.

(d): If $\mathfrak{q} \not\geq \mathfrak{p}$ then $(S/\mathfrak{p})_\mathfrak{q} = 0$ by the argument of (a). If $\mathfrak{q} \geq \mathfrak{p}$ then $K(\mathfrak{p}) = (S/\mathfrak{p})_\mathfrak{p}$ is a further localization of $(S/\mathfrak{p})_\mathfrak{q}$, and $K(\mathfrak{p}) \neq 0$ by (b), so $(S/\mathfrak{p})_\mathfrak{q} \neq 0$. The claim follows, using (b). □

We can now give a number of new characterizations of the category $\mathcal{M}_\mathfrak{p} = \operatorname{loc}\langle K(\mathfrak{p}) \rangle$.

Theorem 6.1.8. *We have* $\operatorname{loc}\langle M_\mathfrak{p} S \rangle = \operatorname{loc}\langle K(\mathfrak{p}) \rangle = \mathcal{M}_\mathfrak{p}$, *and* $\langle M_\mathfrak{p} S \rangle = \langle K(\mathfrak{p}) \rangle$. *Moreover, for any object* $X \in \mathcal{C}$, *the following are equivalent.*

(a) $X = X_\mathfrak{p}$ *and* $X_\mathfrak{q} = 0$ *for all* $\mathfrak{q} < \mathfrak{p}$.
(b) $\pi_*(X)$ *is* $\mathfrak{p}$*-local and* $\mathfrak{p}$*-torsion.*
(c) $X = M_\mathfrak{p} X$.
(d) $\langle X \rangle \leq \langle M_\mathfrak{p} S \rangle = \langle K(\mathfrak{p}) \rangle$.
(e) $X \in \operatorname{loc}\langle M_\mathfrak{p} S \rangle = \operatorname{loc}\langle K(\mathfrak{p}) \rangle = \mathcal{M}_\mathfrak{p}$.

Proof. Consider the following auxiliary statements:

(d′) $\langle X \rangle \leq \langle M_\mathfrak{p} S \rangle$.
(e′) $X \in \operatorname{loc}\langle M_\mathfrak{p} S \rangle$.
(e″) $X \in \operatorname{loc}\langle K(\mathfrak{p}) \rangle$.

We shall prove that

$$(\mathrm{e}'') \Rightarrow (\mathrm{e}') \Rightarrow (\mathrm{c}) \Rightarrow (\mathrm{d}') \Rightarrow (\mathrm{a}) \Rightarrow (\mathrm{b}) \Rightarrow (\mathrm{e}'').$$

This implies everything except that $\langle M_{\mathfrak{p}}S\rangle = \langle K(\mathfrak{p})\rangle$. However, for any U, the category $\{V \mid \langle V\rangle \leq \langle U\rangle\}$ is localizing. Using this and the equality $\mathrm{loc}\langle M_{\mathfrak{p}}S\rangle = \mathrm{loc}\langle K(\mathfrak{p})\rangle$, we easily deduce that $\langle M_{\mathfrak{p}}S\rangle = \langle K(\mathfrak{p})\rangle$.

Before we start, note that any of (a)–(e) (or the primed versions) implies that X is $\mathfrak{p}$-local. Nothing changes if we replace $\mathcal{C}$ by the $\mathfrak{p}$-local category $\mathcal{C}_{\mathfrak{p}}$, so we may assume that $R = \pi_*(S)$ is local with maximal ideal $\mathfrak{p}$. This implies that $K(\mathfrak{p}) = S/\mathfrak{p}$, and that $M_{\mathfrak{p}}$ is just the finite colocalization functor $C_{<\mathfrak{p}}$ defined by the thick subcategory

$$\mathcal{A} = \{Z \text{ small} \mid Z_{\mathfrak{q}} = 0 \text{ for all } \mathfrak{q} < \mathfrak{p}\}.$$

Having made these assumptions, we can drop the explicit $\mathfrak{p}$-local hypotheses in (a) and (b).

(e$''$) $\Rightarrow$ (e$'$): For any $\mathfrak{q} < \mathfrak{p}$ we have $(S/\mathfrak{p})_{\mathfrak{q}} = 0$. Thus $S/\mathfrak{p} \in \mathcal{A}$, so $S/\mathfrak{p} = C_{<\mathfrak{p}}S \wedge S/\mathfrak{p} \in \mathrm{loc}\langle C_{<\mathfrak{p}}S\rangle$.

(e$'$) $\Rightarrow$ (c): If $X \in \mathrm{loc}\langle C_{<\mathfrak{p}}S\rangle$ then $L_{<\mathfrak{p}}X = 0$ so $X = C_{<\mathfrak{p}}X$.

(c) $\Rightarrow$ (d$'$): This is clear, as $C_{<\mathfrak{p}}X = C_{<\mathfrak{p}}S \wedge X$.

(d$'$) $\Rightarrow$ (a): For $\mathfrak{q} < \mathfrak{p}$ we have $L_{\mathfrak{q}}\mathcal{A} = 0$ (by the definition of $\mathcal{A}$) and thus $S_{\mathfrak{q}} \wedge C_{<\mathfrak{p}}S = 0$ (as $C_{<\mathfrak{p}}S \in \mathrm{loc}\langle\mathcal{A}\rangle$). Thus, if $\langle X\rangle \leq \langle C_{<\mathfrak{p}}S\rangle$, then $X_{\mathfrak{q}} = S_{\mathfrak{q}} \wedge X = 0$.

(a) $\Rightarrow$ (b): If (a) holds, then $\pi_*(X)_{\mathfrak{q}} = 0$ for $\mathfrak{q} < \mathfrak{p}$. A well-known piece of algebra implies that $\pi_*(X)$ is then $\mathfrak{p}$-torsion (use the following fact: an element $x \in N$ is nonzero in $N_{\mathfrak{q}}$ if and only if the radical of the annihilator of x is contained in $\mathfrak{q}$).

(b) $\Rightarrow$ (e$''$): Suppose that $\pi_*(X)$ is $\mathfrak{p}$-torsion. The localizing category

$$\mathcal{D} = \{Y \mid \pi_*(Y) \text{ is } \mathfrak{p}\text{-torsion }\}$$

contains X, so it also contains $Y = L^f_{S/\mathfrak{p}}S \wedge X = L^f_{S/\mathfrak{p}}X$. Choose generators $(y_0, \dots, y_{n-1})$ for $\mathfrak{p}$, and set $Y_k = F(S/(y_0, \dots, y_{k-1}), Y) = D(S/(y_0, \dots, y_{k-1})) \wedge Y$. Note that $Y_n = F(S/\mathfrak{p}, L^f_{S/\mathfrak{p}}X) = 0$ by the definition of $L^f_{S/\mathfrak{p}}$. If $Y_k = Y_{k-1}/y_{k-1} = 0$ then y_{k-1} acts isomorphically on $\pi_*(Y_{k-1})$, but it also acts nilpotently as $y_{k-1} \in \mathfrak{p}$ and $Y_{k-1} \in \mathcal{D}$. It follows that $\pi_*(Y_{k-1}) = 0$, and thus that $Y_{k-1} = 0$. By downwards induction, we conclude that $Y = Y_0 = 0$, so $X = C^f_{S/\mathfrak{p}}X \in \mathrm{loc}\langle S/\mathfrak{p}\rangle$ as required. □

Theorem 6.1.9. *We have an equality of Bousfield classes*

$$\langle S\rangle = \coprod_{\mathfrak{p}\in \mathrm{Spec}\, R} \langle K(\mathfrak{p})\rangle.$$

(*Note also that* $\langle K(\mathfrak{p})\rangle \wedge \langle K(\mathfrak{q})\rangle = 0$ *for* $\mathfrak{p} \neq \mathfrak{q}$, *by Proposition* 6.1.7.)

Proof. Suppose that $K(\mathfrak{p}) \wedge X = 0$ for all $\mathfrak{p}$; we need to show that $X = 0$. By Theorem 6.1.8, we have $\langle M_{\mathfrak{p}}\rangle = \langle K(\mathfrak{p})\rangle$, so $M_{\mathfrak{p}}X = 0$ for all $\mathfrak{p}$. We claim that $X_{\mathfrak{p}} = 0$ for all $\mathfrak{p}$; we prove this by induction on $\mathfrak{p}$. Fix $\mathfrak{p}$ and suppose that $X_{\mathfrak{q}} = 0$ for all $\mathfrak{q} < \mathfrak{p}$. Then by the implication (a)$\Rightarrow$(c) of Theorem 6.1.8 (applied to $X_{\mathfrak{p}}$), we find that $X_{\mathfrak{p}} = M_{\mathfrak{p}}X = 0$. Thus $X_{\mathfrak{p}} = 0$ for all $\mathfrak{p}$; using Lemma 6.0.3, we conclude that $\pi_*(X) = 0$ and thus $X = 0$. □

It follows that the nilpotence results of Section 5 apply:

Corollary 6.1.10. *Let $\mathcal{C}$ be a Noetherian stable homotopy category. Then the objects $\{K(\mathfrak{p}) \mid \mathfrak{p} \in \mathrm{Spec}\, R\}$ detect nilpotence, in the sense that Theorem* 5.1.3 *applies. Moreover, Theorem* 5.1.2 *applies with $K(\mathfrak{p})$ replaced by $M_{\mathfrak{p}}$.* □

We pause briefly to deduce the claimed relation between our two conjectures.

Proposition 6.1.11. *If $\mathcal{M}_{\mathfrak{p}}$ is minimal among nontrivial localizing subcategories, then $\langle K(\mathfrak{p})\rangle$ is a minimal Bousfield class.*

Proof. This now follows from Proposition 3.7.4, Theorem 6.1.9, and part (a) of Proposition 6.1.7. □

6.2. **Thick subcategories.** We next attempt to classify thick subcategories of small objects in a Noetherian stable homotopy category $\mathcal{C}$; we succeed completely if each $\langle K(\mathfrak{p})\rangle$ is a minimal Bousfield class. We first recall some convenient terminology.

Definition 6.2.1. Let $\mathcal{A} \xrightarrow{f} \mathcal{B} \xrightarrow{g} \mathcal{A}$ be maps of partially ordered sets. We say that g is *left adjoint* to f (and write $g \vdash f$) if $g(b) \leq a$ is equivalent to $b \leq f(a)$. (It is equivalent to say that g is left adjoint to f when $\mathcal{A}$ and $\mathcal{B}$ are regarded as categories, and f and g as functors, in the usual way.)

The particular lattices of interest are as follows.

Definition 6.2.2. We define

$$\mathcal{L}_f^{\mathrm{op}} = \{ \text{ thick subcategories of small objects } \},$$
$$\mathcal{L}_t = \{ \text{ subsets } T \subseteq \operatorname{Spec} R \text{ that are closed under specialization } \}.$$

Note that $\mathcal{L}_f^{\mathrm{op}}$ is antiisomorphic to the lattice of finite localization functors, by Proposition 3.8.3. We define $f\colon \mathcal{L}_t \longrightarrow \mathcal{L}_f^{\mathrm{op}}$ by

$$f(T) = \operatorname{thick}\langle S/\mathfrak{p} \mid \mathfrak{p} \in T\rangle = \{Z \in \mathcal{F} \mid \operatorname{supp}(Z) \subseteq T\}.$$

(We shall verify below that these two definitions are equivalent.) We also define maps $g, g'\colon \mathcal{L}_f^{\mathrm{op}} \longrightarrow \mathcal{L}_t$ by

$$g(\mathcal{A}) = \operatorname{supp}(\mathcal{A}), \qquad g'(\mathcal{A}) = \{\mathfrak{p} \mid S/\mathfrak{p} \in \mathcal{A}\}.$$

(We shall verify below that these sets are closed under specialization).

Theorem 6.2.3. *The two definitions of f given above are the same, and the sets $g(\mathcal{A})$ and $g'(\mathcal{A})$ are closed under specialization. Moreover, we have*

$$g \vdash f \vdash g', \qquad g \leq g',$$
$$gf = 1 = g'f, \qquad fg' \leq 1 \leq fg.$$

If each $\langle K(\mathfrak{p})\rangle$ is a minimal Bousfield class, then f, g and g' are isomorphisms with $g = g' = f^{-1}$.

Proof. We first show that the two definitions of f are equivalent. Consider a set $T \in \mathcal{L}_t$, and write $\mathcal{A} = \operatorname{thick}\langle S/\mathfrak{p} \mid \mathfrak{p} \in T\rangle$ and $\mathcal{B} = \{Z \in \mathcal{F} \mid \operatorname{supp}(Z) \subseteq T\}$. If $\mathfrak{p} \in T$ then $\operatorname{supp}(S/\mathfrak{p}) = V(\mathfrak{p}) \subseteq T$ (because T is closed under specialization). It follows that $\mathcal{A} \subseteq \mathcal{B}$. For the converse, write $C = C_{\mathcal{A}}^f$ and $L = L_{\mathcal{A}}^f$. Suppose that $Z \in \mathcal{B}$. If $\mathfrak{p} \notin T$ we have $Z \wedge K(\mathfrak{p}) = 0$ (by the definition of $\mathcal{B}$) and thus $LZ \wedge K(\mathfrak{p}) = LS \wedge Z \wedge K(\mathfrak{p}) = 0$. On the other hand, if $\mathfrak{p} \in T$ then we have $S/\mathfrak{p} \in \mathcal{A}$, so $LS/\mathfrak{p} = 0$ (by the definition of L), so $LZ \wedge K(\mathfrak{p}) = Z \wedge S_{\mathfrak{p}} \wedge LS/\mathfrak{p} = 0$. It follows that $LZ \wedge K(\mathfrak{p}) = 0$ for all $\mathfrak{p}$, so $LZ = 0$ by Theorem 6.1.9. Theorem 3.3.3 now tells us that $Z \in \mathcal{A}$. Thus $\mathcal{A} = \mathcal{B}$ as claimed.

We next show that $g(\mathcal{A})$ and $g'(\mathcal{A})$ are closed under specialization. For $g(\mathcal{A}) = \operatorname{supp}(\mathcal{A})$, this is immediate from Proposition 6.1.7. For $g'(\mathcal{A})$, it is immediate from Lemma 6.0.9.

Using our first definition of f, we see that $T \subseteq g'(\mathcal{A})$ if and only if $f(T) \subseteq \mathcal{A}$, so that $f \vdash g'$. Using the second definition, it is clear that $g(\mathcal{A}) \subseteq T$ if and only if $\mathcal{A} \subseteq f(T)$, so that $g \vdash f$. It is clear that $g' \leq g$. By setting $T = g(\mathcal{A})$ or $T = g'(\mathcal{A})$ or $\mathcal{A} = f(T)$ in the adjunctions above, we obtain (co)unit inequalities

$$fg'(\mathcal{A}) \subseteq \mathcal{A} \subseteq fg(\mathcal{A})$$

$$gf(T) \leq T \leq g'f(T)$$

By combining the latter with the inequality $g' \leq g$, we conclude that $gf(T) = T = g'f(T)$.

Now suppose that all the Bousfield classes $\langle K(\mathfrak{p})\rangle$ are minimal. We can then apply Theorem 6.1.9 and Corollary 5.2.3 to show that the collection of $K(\mathfrak{p})$'s determines thick subcategories, in other words that the map g is injective. As $gf = 1$, we conclude that g is also surjective, in fact that f and g are inverse isomorphisms. As $g'f = 1$, we see that $g' = f^{-1} = g$. □

6.3. **Localizing subcategories.** We next study localizing subcategories of $\mathcal{C}$. We start with the following lemma.

Lemma 6.3.1. *Suppose that $T \subseteq \operatorname{Spec} R$ is closed under specialization, and let $\mathcal{A} = f(T) = \operatorname{thick}\langle S/\mathfrak{p} \mid \mathfrak{p} \in T\rangle$. Then*

$$C^f_{\mathcal{A}} K(\mathfrak{p}) = \begin{cases} K(\mathfrak{p}) & \text{if } \mathfrak{p} \in T, \\ 0 & \text{otherwise.} \end{cases}$$

Proof. If $\mathfrak{p} \in T$ then $S/\mathfrak{p} \in \mathcal{A}$, so $K(\mathfrak{p}) \in \operatorname{loc}\langle S/\mathfrak{p}\rangle \subseteq \operatorname{loc}\langle \mathcal{A}\rangle$, so $C^f_{\mathcal{A}} K(\mathfrak{p}) = K(\mathfrak{p})$. On the other hand, if $\mathfrak{p} \notin T = \operatorname{supp}(\mathcal{A})$ then for all $Z \in \mathcal{A}$ we have $Z_{\mathfrak{p}} = 0$, so $[Z, K(\mathfrak{p})] = [Z_{\mathfrak{p}}, K(\mathfrak{p})] = 0$. It follows that $C^f_{\mathcal{A}} K(\mathfrak{p}) = 0$. □

The key fact for our understanding of localizing subcategories is as follows.

Proposition 6.3.2. *For any object $X \in \mathcal{C}$ we have*

$$X \in \operatorname{loc}\langle K(\mathfrak{p}) \wedge X \mid \mathfrak{p} \in \operatorname{Spec} R\rangle.$$

Proof. Write $\mathcal{D} = \operatorname{loc}\langle K(\mathfrak{p}) \wedge X \mid \mathfrak{p} \in \operatorname{Spec} R\rangle$, which is the same as $\operatorname{loc}\langle M_{\mathfrak{p}} X \mid \mathfrak{p} \in \operatorname{Spec} R\rangle$, because $\operatorname{loc}\langle M_{\mathfrak{p}} S\rangle = \operatorname{loc}\langle K(\mathfrak{p})\rangle$. Write $T = \{\mathfrak{p} \in \operatorname{Spec} R \mid S/\mathfrak{p} \wedge X \in \mathcal{D}\}$; this is closed under specialization by Lemma 6.0.9. Define $\mathcal{A} = f(T) = \operatorname{thick}\langle S/\mathfrak{p} \mid \mathfrak{p} \in T\rangle$ and $C_T = C^f_{\mathcal{A}}$. Note that $C_T X \in \operatorname{loc}\langle S/\mathfrak{p} \wedge X \mid \mathfrak{p} \in T\rangle \subseteq \mathcal{D}$, by the definition of T.

We claim that $T = \operatorname{Spec} R$. If not, then as R is Noetherian, we can choose a maximal element $\mathfrak{p}$ of $\operatorname{Spec} R \setminus T$. Maximality means that $T' = T \cup \{\mathfrak{p}\}$ is closed under specialization. There is a morphism $C_T \longrightarrow C_{T'}$ of colocalization functors; call the cofiber M. By Lemma 6.3.1, we have $C_T K(\mathfrak{q}) = C_{T'} K(\mathfrak{q})$ (and thus $MS \wedge K(\mathfrak{q}) = MK(\mathfrak{q}) = 0$) unless $\mathfrak{q} = \mathfrak{p}$. It follows from Theorem 6.1.9 that $\langle MS\rangle \leq \langle K(\mathfrak{p})\rangle$, and thus from Theorem 6.1.8 that $MS \in \operatorname{loc}\langle M_{\mathfrak{p}} S\rangle$, and thus that $MX \in \operatorname{loc}\langle M_{\mathfrak{p}} X\rangle \subseteq \mathcal{D}$ (by the definition of $\mathcal{D}$). We observed above that $C_T X \in \mathcal{D}$, so we conclude from the cofibration that $C_{T'} X \in \mathcal{D}$. As $S/\mathfrak{p} \wedge C_{T'} S = C_{T'} S/\mathfrak{p} = S/\mathfrak{p}$, we conclude that $S/\mathfrak{p} \wedge X \in \mathcal{D}$, contradicting our assumption that $\mathfrak{p} \notin T$. Thus $T = \operatorname{Spec} R$ as claimed.

We thus have $\mathcal{A} = f(\operatorname{Spec} R) = \{Z \mid \operatorname{supp}(Z) \subseteq \operatorname{Spec} R\} = \mathcal{F}$, so $X = C_T X \in \mathcal{D}$, as required. □

We can deduce the following splitting of the Bousfield lattice.

Corollary 6.3.3. *Let $\mathcal{B}_\mathfrak{p}$ be the lattice of localizing subcategories that are contained in $\mathcal{M}_\mathfrak{p}$, and $\mathcal{B}$ the lattice of all localizing subcategories. Then there is a natural isomorphism $\mathcal{B} = \prod_\mathfrak{p} \mathcal{B}_\mathfrak{p}$.*

Proof. The map $\mathcal{B} \longrightarrow \prod_\mathfrak{p} \mathcal{B}_\mathfrak{p}$ sends $\mathcal{D}$ to the collection of subcategories $\mathcal{D}_\mathfrak{p} = \mathcal{D} \cap \mathcal{M}_\mathfrak{p}$. The map the other way sends a collection of subcategories $\mathcal{E}_\mathfrak{p}$ to $\mathcal{E} = \operatorname{loc}\langle \bigcup_\mathfrak{p} \mathcal{E}_\mathfrak{p} \rangle$. It is clear that $\mathcal{E}_\mathfrak{p} \subseteq \mathcal{E} \cap \mathcal{M}_\mathfrak{p}$. Conversely, if $X \in \mathcal{E} \cap \mathcal{M}_\mathfrak{p}$ then

$$X = M_\mathfrak{p} S \wedge X \in \operatorname{loc}\langle \bigcup_\mathfrak{q} M_\mathfrak{p} S \wedge \mathcal{E}_\mathfrak{q} \rangle = M_\mathfrak{p} S \wedge \mathcal{E}_\mathfrak{p} = \mathcal{E}_\mathfrak{p},$$

so $\mathcal{E} \cap \mathcal{M}_\mathfrak{p} = \mathcal{E}_\mathfrak{p}$. It is also clear that $\operatorname{loc}\langle \bigcup_\mathfrak{p} \mathcal{D}_\mathfrak{p} \rangle \subseteq \mathcal{D}$, and Proposition 6.3.2 implies the opposite inequality. It follows that these two constructions are mutually inverse, as required. □

Corollary 6.3.4. *If Conjecture* 6.1.2 *holds for $\mathcal{C}$, then every localizing subcategory of $\mathcal{C}$ is closed, and the lattice of such subcategories is isomorphic to the lattice of subsets of* $\operatorname{Spec} R$ *(and antiisomorphic to the Bousfield lattice).*

Proof. Conjecture 6.1.2 says that each $\mathcal{M}_\mathfrak{p}$ is minimal, so that each lattice $\mathcal{B}_\mathfrak{p}$ is isomorphic to the two-element lattice $\{0, 1\}$, so that $\mathcal{B} = \prod_{\mathfrak{p} \in \operatorname{Spec} R} \mathcal{B}_\mathfrak{p}$ is isomorphic to the lattice of subsets of $\operatorname{Spec} R$. We can describe the maps more explicitly. Let T be a subset of $\operatorname{Spec} R$, and write $X = \coprod_{\mathfrak{q} \notin T} K(\mathfrak{q})$. Then the corresponding localizing subcategory is

$$\operatorname{loc}\langle K(\mathfrak{p}) \mid \mathfrak{p} \in T \rangle = \{Y \mid X \wedge Y = 0\}.$$

It follows from Lemma 3.6.6 that this is a closed localizing subcategory; thus all localizing subcategories are closed. In any stable homotopy category, the Bousfield lattice is antiisomorphic to the category of closed localizing subcategories. □

We now turn to the telescope conjecture, in other words, the classification of smashing localization functors. For this, we need to study the objects $S_\mathfrak{p}^\diamond$.

Lemma 6.3.5.

(a) *There is a natural isomorphism $\pi_*(S_\mathfrak{p}^\diamond \wedge X) = R_\mathfrak{p}^\diamond \otimes_{R_\mathfrak{p}} \pi_*(X_\mathfrak{p})$, and this is the same as $\pi_*(X)_\mathfrak{p}^\diamond$ if $\pi_*(X_\mathfrak{p})$ is finitely generated over $R_\mathfrak{p}$.*
(b) $\langle S_\mathfrak{p}^\diamond \rangle = \langle S_\mathfrak{p} \rangle = \coprod_{\mathfrak{q} \leq \mathfrak{p}} \langle K(\mathfrak{q}) \rangle$.
(c) $L_{K(\mathfrak{p})} S = S_\mathfrak{p}^\diamond$.

Proof. (a): There is an obvious pairing $\pi_*(Y) \otimes_R \pi_*(X) \longrightarrow \pi_*(Y \wedge X)$. Taking $Y = S_\mathfrak{p}^\diamond$ gives a map $R_\mathfrak{p}^\diamond \otimes_R \pi_*(X) \longrightarrow \pi_*(S_\mathfrak{p}^\diamond \wedge X)$. As $S_\mathfrak{p}^\diamond$ is $\mathfrak{p}$-local, the left hand side is the same as $R_\mathfrak{p}^\diamond \otimes_{R_\mathfrak{p}} \pi_*(X_\mathfrak{p})$. This is a homology functor of X, because $R_\mathfrak{p}^\diamond$ is flat over $R_\mathfrak{p}$, and $\pi_*(S_\mathfrak{p}^\diamond \wedge X)$ is also a homology functor for more obvious reasons. We thus have a map of graded homology functors that is an isomorphism when $X = S$, and thus for all X. It is well-known that $R_\mathfrak{p}^\diamond \otimes_{R_\mathfrak{p}} N = N_\mathfrak{p}^\diamond$ for finitely generated modules over $R_\mathfrak{p}$, which gives the last statement.

(b): Using (a) and the fact that $R_\mathfrak{p}^\diamond$ is *faithfully* flat over $R_\mathfrak{p}$, we see that $\langle S_\mathfrak{p}^\diamond \rangle = \langle S_\mathfrak{p} \rangle$. By smashing the equivalence $\langle S \rangle = \coprod_\mathfrak{q} \langle K(\mathfrak{q}) \rangle$ with $S_\mathfrak{p}$, we see that this is the same as $\coprod_{\mathfrak{q} \leq \mathfrak{p}} \langle K(\mathfrak{q}) \rangle$.

(c): Observe that $\pi_* K(\mathfrak{p})$ is a $\mathfrak{p}$-torsion module, finitely generated over $R_\mathfrak{p}$, so that $\pi_*(S_\mathfrak{p}^\diamond \wedge K(\mathfrak{p})) = \pi_*(K(\mathfrak{p}))_\mathfrak{p}^\diamond = \pi_*(K(\mathfrak{p}))$ by (a). Thus the obvious map $S \longrightarrow S_\mathfrak{p}^\diamond$ gives an equivalence $S \wedge K(\mathfrak{p}) = S_\mathfrak{p}^\diamond \wedge K(\mathfrak{p})$ and thus $L_{K(\mathfrak{p})}S = L_{K(\mathfrak{p})}S_\mathfrak{p}^\diamond$. It will therefore be enough to show that $S_\mathfrak{p}^\diamond$ is $K(\mathfrak{p})$-local. Suppose that $K(\mathfrak{p}) \wedge X = 0$, so we need to prove that $[X, S_\mathfrak{p}^\diamond] = 0$. Recall that $S_\mathfrak{p}^\diamond = F(I_\mathfrak{p}, I_\mathfrak{p})$, so it is enough to show that $X \wedge I_\mathfrak{p} = 0$. However, $\pi_* I_\mathfrak{p}$ is $\mathfrak{p}$-local and $\mathfrak{p}$-torsion, so Theorem 6.1.8 tells us that $\langle I_\mathfrak{p} \rangle \leq \langle K(\mathfrak{p}) \rangle$, and $K(\mathfrak{p}) \wedge X = 0$, so $I_\mathfrak{p} \wedge X = 0$ as required. □

Corollary 6.3.6. *If L is a smashing localization functor and $\langle LS \rangle \geq \langle K(\mathfrak{p}) \rangle$, then $\langle LS \rangle \geq \coprod_{\mathfrak{q} \leq \mathfrak{p}} \langle K(\mathfrak{q}) \rangle$.*

Proof. If $\langle K(\mathfrak{p}) \rangle \leq \langle LS \rangle$, then $L_{K(\mathfrak{p})}S$ is L_{LS}-local, which is the same as being L-local, as L is smashing; in other words $LS \wedge L_{K(\mathfrak{p})}S = L_{K(\mathfrak{p})}S$. It follows that

$$\coprod_{\mathfrak{q} \leq \mathfrak{p}} \langle K(\mathfrak{q}) \rangle = \langle L_{K(\mathfrak{p})}S \rangle = \langle LS \wedge L_{K(\mathfrak{p})}S \rangle \leq \langle LS \rangle.$$

□

We can now prove the promised classification theorem.

Theorem 6.3.7. *Suppose that each Bousfield class $\langle K(\mathfrak{p}) \rangle$ is minimal. Then the telescope conjecture holds for $\mathcal{C}$—every smashing localization is a finite localization. The lattice of finite localizations is antiisomorphic to the lattice of thick subcategories of small objects, which is isomorphic to the lattice of subsets of* Spec R *that are closed under specialization.*

Proof. Let L be a smashing localization functor, and write

$$T = \{\mathfrak{p} \in \operatorname{Spec} R \mid LK(\mathfrak{p}) = 0\}.$$

If $LS \wedge K(\mathfrak{p}) \neq 0$ then $\langle LS \rangle \geq \langle K(\mathfrak{p}) \rangle$ by minimality, and thus $LS \wedge K(\mathfrak{q}) \neq 0$ for $\mathfrak{q} \leq \mathfrak{p}$ by Corollary 6.3.6. It follows that T is closed under specialization. Let L_T be the corresponding localization functor. By Lemma 6.3.1, we have $L_T K(\mathfrak{p}) = 0$ if and only if $\mathfrak{p} \in T$, and $L_T K(\mathfrak{p}) = K(\mathfrak{p})$ otherwise. By smashing the decomposition $\langle S \rangle = \coprod_\mathfrak{p} \langle K(\mathfrak{p}) \rangle$ with LS and $L_T S$, we see that $\langle LS \rangle = \langle L_T S \rangle$. As a localization functor is determined by the corresponding category of local objects, we have $L = L_T$. Thus, every smashing localization is a finite localization. The rest follows from Proposition 6.2.3 and Proposition 3.8.3. □

7. Connective stable homotopy theory

In this section, we discuss stable homotopy categories with a good notion of connectivity. These share many features with the homotopy category of spectra.

Definition 7.1.1. Suppose that $\mathcal{C}$ is a monogenic stable homotopy category. We say that $\mathcal{C}$ is *connective* if $\pi_n S = 0$ for $n < 0$.

Of course, this is the case for ordinary stable homotopy theory. In fact, it is the case for many of our examples in Section 1.2 (including $\mathcal{D}(R)$ and $\mathcal{C}(B)$). However, Bousfield localizations of connective categories are rarely connective. Here we do little more than summarize the properties of connective categories and briefly sketch a proof, referring to the work of Margolis for details.

Proposition 7.1.2. *Let $\mathcal{C}$ be a connective monogenic stable homotopy category.*

(a) *Suppose that* $X \in \mathcal{C}$ *has* $\pi_m X = 0$ *for* $m < k$ *(in this case, we say* X *is* bounded below*). Then there is a cellular tower*

$$X_k \xrightarrow{f_k} X_{k+1} \xrightarrow{f_{k+1}} X_{k+2} \xrightarrow{f_{k+2}} \cdots$$

whose minimal weak colimit is X*, such that* X_k *is a coproduct of copies of* S^k*, and such that the cofiber of* f_n *is a coproduct of copies of* S^{n+1}*.*

(b) *Suppose that* $R = \pi_0 S$ *is a Noetherian ring of finite global dimension, and that every projective* R*-module is free. Then every small object of* $\mathcal{C}$ *has a finite cellular tower as in* (a)*.*

(c) *Given an object* $X \in \mathcal{C}$ *and an integer* k*, there is a diagram* $X[k,\infty] \xrightarrow{f} X \xrightarrow{g} X[-\infty, k-1]$ *such that* $\pi_m X[k,\infty] = 0$ *for* $m < k$*,* $\pi_m X[-\infty, k-1] = 0$ *for* $m \geq k$*,* $\pi_m f$ *is an isomorphism for* $m \geq k$*, and* $\pi_m g$ *is an isomorphism for* $m < k$*. Furthermore,* $[X[k,\infty], X[-\infty, k-1]] = 0$*. In the terminology of* [BBD82]*,* $\mathcal{C}$ *admits a* t-structure.

(d) *Using the diagram in part* (c)*, we can construct a* Postnikov tower *for any object* X*:*

$$\begin{array}{ccccccccc}
\cdots & \longrightarrow & X[-\infty, r] & \longrightarrow & X[-\infty, r-1] & \longrightarrow & X[-\infty, r-2] & \longrightarrow & \cdots \\
 & & \uparrow & & \uparrow & & \uparrow & & \\
 & & X[r] & & X[r-1] & & X[r-2] & &
\end{array}$$

The sequential colimit of $X[-\infty, r]$ *is* 0 *and the sequential limit of* $X[-\infty, r]$ *is* X*. (See* [Mar83, Chapter 5].)

(e) *Let* R *denote the ring* $\pi_0 S$*. Let* $\mathcal{A}$ *denote the full subcategory of* $\mathcal{C}$ *consisting of objects* X *such that* $\pi_m X = 0$ *unless* $m = 0$*. Then* $\mathcal{A}$ *is a closed symmetric monoidal Abelian category, and is equivalent as such to the category of* R*-modules. In the terminology of* [BBD82]*, the category of* R*-modules forms the* heart *of* $\mathcal{C}$*. We call elements of* $\mathcal{A}$ Eilenberg-MacLane objects.

(f) *Let* $H = S[0]$ *denote the object of* $\mathcal{A}$ *corresponding to* R*. We call* $H_* X = \pi_*(H \wedge X)$ *the* ordinary homology *of* X*. Then* H *is a ring object, and if* $\pi_m X = 0$ *for all* $m < k$*, then the natural (Hurewicz) map* $\pi_k X \longrightarrow H_k X$ *is an isomorphism.*

Proof. Given an object X and an integer k, we construct $X[-\infty, k]$ as follows. We set $X_0 = X$, choose a system of generators $\{x_i\}$ for $\pi_{k+1} X_0$ as a module over R, let B_1 be a coproduct of copies of S^{k+1} (one for each i), and consider the evident map $B_1 \longrightarrow X_0$. We define X_1 to be the cofiber of this map. One can check that $\pi_m X_1 = \pi_m X$ for $m \leq k$, and $\pi_k X_1 = 0$. In a similar way, one can construct a coproduct B_2 of copies of S^{k+2} and a cofibration $B_2 \longrightarrow X_1 \longrightarrow X_2$, such that $\pi_* X_2 = \pi_* X$ below degree k, and $\pi_{k+1} X_2 = \pi_{k+2} X_2 = 0$. Continuing in this manner and passing to the sequential colimit, we get $X_\infty = X[-\infty, k]$. This comes equipped with a natural map $X \longrightarrow X[-\infty, k]$, and we define $X[k+1, \infty]$ to be the fiber. We also define $X[k,l] = X[k,\infty][-\infty, l]$ and $X[k] = X[k,k]$.

The proof of (a) is essentially identical to the construction above. We refer to [Mar83, Chapter 5] for the proofs of (c), (d) and (f). For (e), observe that π_0 is a functor from $\mathcal{A}$ to the category $\mathcal{M}$ of R-modules. This is essentially surjective: for any module M, we can choose a presentation $\bigoplus_i R \to \bigoplus_j R \to M$, construct a corresponding cofiber sequence $\coprod_i S \to \coprod_j S \to X$, and then $X[0] \in \mathcal{A}$ and

$\pi_0 X[0] = M$. Small modifications of the arguments of Margolis show that this is also full and faithful.

For (b), suppose that R is Noetherian with finite global dimension, and that all projective R-modules are free. Let X be a small object in $\mathcal{C}$. It is easy to see that $H_* X$ is then a finitely generated graded module over R, and that $\pi_k X = 0$ for $k \ll 0$. We may assume without loss of generality that $\pi_k X = 0$ for $k < 0$. Write $X^0 = X$. Much as above, we let B_0 be a finite coproduct of copies of S^0 and choose a map $B_0 \longrightarrow X$ that is surjective on π_0 (which agrees with H_0 by (f)). We let X^1 be the cofiber, and note that $H_* X^1$ is the same as $H_* X^0$ except in dimension zero (where $H_0 X^1 = 0$) and dimension one. Continuing in this way, we get a sequence of cofibrations $B_k \longrightarrow X^k \longrightarrow X^{k+1}$, where $H_* X^k$ is zero below dimension k and agrees with $H_* X^{k-1}$ above dimension k. It follows that for large k, the groups $H_* X^k$ are concentrated in a single degree. After that point, the projective dimension of the single group in question decreases by one at each stage, until it becomes zero, so the group is free. At that point, we can choose the map $B_k \longrightarrow X^k$ so that it induces an isomorphism in homology, so that $H_* X^{k+1} = 0$. As $\pi_* X^{k+1} = 0$ for $k \ll 0$, part (f) implies that $\pi_* X^{k+1} = 0$, and thus $X^{k+1} = 0$. Now let X_j be the fiber of the evident map $X \longrightarrow X^j$, to get a finite cellular tower of the required type. □

8. Semisimple stable homotopy theory

In this section we give conditions under which a stable homotopy category is actually an Abelian category. The most familiar example is the category of rational spectra, which is equivalent to the category of graded rational vector spaces. We will allow $\mathcal{C}$ to be multigraded as in Section 1.3.

Definition 8.1.1. Suppose that $\mathcal{C}$ is an algebraic stable homotopy category. We say that $\mathcal{C}$ is *semisimple* if, for every pair $Y, Z \in \mathcal{G}$, we have

(a) If $Y \neq Z$, then $[Y, Z]_* = 0$; and
(b) $[Z, Z]_*$ is a (multigraded) division algebra k_Z, where the multiplication is given by composition.

Example 8.1.2. Consider the category $\mathcal{C}(kG)$ of chain complexes of projective kG-modules where G is a finite group and $p = \text{char}(k)$ does not divide $|G|$. Then kG is semisimple, so every kG-module is a direct sum of simple modules. Also, every simple module appears as a summand of kG; hence every kG-module is projective. Schur's lemma says that if S and T are non-isomorphic simple kG-modules, then $\text{Hom}_{kG}(S, T) = 0$, while $k_S = \text{Hom}_{kG}(S, S)$ is a division algebra.

In this case we are doing something just a bit more complicated than rational stable homotopy (equivalently, graded rational linear algebra).

Proposition 8.1.3. *If $\mathcal{C}$ is a semisimple stable homotopy category , then every object X of $\mathcal{C}$ is equivalent to a coproduct of suspensions of elements of $\mathcal{G}$.*

Proof. Given an object X of $\mathcal{C}$ and $Z \in \mathcal{G}$, note that $[Z, X]_*$ is a right module over the division algebra k_Z; let B_Z denote a basis. We have a map F given by

$$\coprod_{Z \in \mathcal{G}} \coprod_{\substack{f \in B_Z \\ |f| = n}} \Sigma^n Z \xrightarrow{F} X.$$

Here n could be a multi-index. We claim that F is an equivalence. It suffices to show that $[Y,F]_*$ is an isomorphism for each $Y \in \mathcal{G}$. This is clear, though:

$$(8.1.1) \quad [Y, \coprod_Z \coprod_{f\in B_Z} \Sigma^{|f|}Z]_* = [Y, \coprod_{f\in B_Y} \Sigma^{|f|}Y]_* = \bigoplus_{f\in B_Y} \Sigma^{-|f|}[Y,Y]_* = \bigoplus_{f\in B_Y} \Sigma^{-|f|}k_Y = [Y,X]_*.$$

□

We can extend this proposition a little as follows.

Definition 8.1.4. Given a semisimple stable homotopy category $\mathcal{C}$, a $\mathcal{G}$-*module* is an assignment of a graded right k_Z-module to each $Z \in \mathcal{G}$. The class of $\mathcal{G}$-modules forms an Abelian category $\mathcal{G}$-Mod in the obvious way, where a sequence is exact if and only if it is exact for each Z.

There is a natural functor F from $\mathcal{C}$ to $\mathcal{G}$-Mod that assigns to X and Z the k_Z-module $[Z,X]_*$.

Proposition 8.1.5. *The functor F is an equivalence of categories.* □

The proof is clear. Note that the induced triangulation on $\mathcal{G}$-Mod is a very simple one. Given a $f\colon M \longrightarrow N$ and a generator Z, the cofiber of f at Z is the direct sum of the suspension of the kernel of f and the cokernel of f as a k_Z-vector space.

So, for example, the set of maps between finite objects W and X is in one-to-one correspondence with matrices of the appropriate shape—if r_{ZW} is the k_Z-rank of $[Z,W]_*$ (and similarly for r_{ZX}), then

$$[W,X]_* = \bigoplus_{Z\in\mathcal{G}} M_{r_{ZX}\times r_{ZW}}(k_Z)$$

(where $M_{r\times s}(k)$ is the set of $r \times s$ matrices with coefficients in k).

Note that one can define the *product* of two stable homotopy categories $\mathcal{C}$ and $\mathcal{C}'$ in the obvious way, and this again gives a stable homotopy category: the objects are pairs (X,X'), and all of the structure in the axioms is defined coordinate-wise. For example $(X,X') \wedge (Y,Y')$ is defined to be $(X \wedge Y, X' \wedge Y')$. We point out that a semisimple stable homotopy category will *not* in general be decomposable into a product of stable homotopy categories, one for each element of $\mathcal{G}$. In the case of $\mathcal{C}(kG)$ where p does not divide $|G|$, if Z is a simple kG-module, then $Z \otimes Z$ does not necessarily decompose as a direct sum of copies of Z (as it would if the category split). Similarly, the function object in $\mathcal{C}(kG)$ will not behave "coordinate-wise," as it would in a product stable homotopy category. Thus, the induced symmetric monoidal structure on $\mathcal{G}$-Mod might be complicated. Indeed, this is the content of classical representation theory.

One can describe the $\mathcal{G}$-ideals in a semisimple stable homotopy category in the following way. Draw a graph with one vertex for each generator, and an edge joining Z to W if and only if $Z \wedge W \neq 0$, and let T be the set of components of this graph. One can check that the $\mathcal{G}$-ideals biject with the subsets of T. In the case of a group algebra, T bijects with the set of blocks of kG.

9. Examples of stable homotopy categories

9.1. **A general method.** Suppose that one wants to do homotopy theory in a category $\mathcal{C}$, which has a natural notion of homotopy on its morphism sets. The first step is to consider the category $h\mathcal{C}$ in which the morphisms are homotopy classes of maps. It is by now familiar that this is insufficient to give a good theory. One must instead formally invert a suitable class of weak equivalences to get a category called $\overline{h}\mathcal{C}$. This procedure cures such pathologies as the long line, which is a non-contractible topological space whose homotopy groups all vanish. However, there is no guarantee *a priori* that the morphism sets $\overline{h}\mathcal{C}(X, Y)$ are actually sets rather than proper classes. In many cases it can be shown that $\overline{h}\mathcal{C}$ is equivalent to some full subcategory $\mathcal{D} \subset h\mathcal{C}$. If $\mathcal{C}$ is a closed model category [Qui67, DS95] we can take $\mathcal{D}$ to be the subcategory of objects which are both cofibrant and fibrant. In some other cases one can use a category of cell objects. This general approach has been much used by May and his coauthors [LMS86] in more recent work [EKMM95, KM95] they have also given closed model structures.

We now state a theorem which codifies these ideas. See Definitions 1.1.1, 1.1.4, 1.1.6, and 3.4.1 for the relevant terms.

Theorem 9.1.1 (Cellular approximation). *Let $\mathcal{C}$ be an enriched triangulated category , and $\mathcal{G}$ a set of small strongly dualizable objects containing the unit S. Let $\mathcal{D}$ denote the localizing subcategory generated by $\mathcal{G}$. Suppose in addition that if $X, Y \in \mathcal{G}$, then $X \wedge Y$ and DX are in $\mathcal{D}$. Then:*

(a) *$\mathcal{D}$ is a unital algebraic stable homotopy category with generating set $\mathcal{G}$.*
(b) *The inclusion functor $J\colon \mathcal{D} \longrightarrow \mathcal{C}$ is a geometric morphism, with right adjoint C say.*
(c) *The functor C preserves coproducts and the unit.*
(d) *Let $\mathcal{S}$ be the class of morphisms $f\colon X \longrightarrow Y$ in $\mathcal{C}$ such that $f_*\colon [Z, X]_* \simeq [Z, Y]_*$ for all $Z \in \mathcal{G}$. Then C induces an equivalence $\mathcal{C}[\mathcal{S}^{-1}] \simeq \mathcal{D}$.*

Note that if $\mathcal{C}$ is an enriched triangulated category in which S is small, then we can take $\mathcal{G} = \{S\}$ and the hypotheses of the theorem are satisfied. In particular, in any unital algebraic stable homotopy category, the localizing subcategory $\operatorname{loc}\langle S\rangle$ is a monogenic stable homotopy category. One might wonder why we need more general sets $\mathcal{G}$ at all. A good example of this is provided by the homotopy category of G-spectra discussed in Section 9.4. There, if we just take $\mathcal{G} = \{S\}$ we only get spectra on which G acts trivially, which is equivalent to the homotopy category of non-equivariant spectra. Therefore, in that case, to get anything new we must take a larger set $\mathcal{G}$.

Proof of Theorem 9.1.1. Let X be an object of $\mathcal{C}$. As in the proof of Theorem 2.3.2, we construct a cofibration $CX \xrightarrow{q} X \xrightarrow{i} LX$ with $CX \in \mathcal{D}$, and $[Z, LX]_* = 0$ for all $Z \in \mathcal{G}$ (and therefore all $Z \in \mathcal{D}$). It follows as in Lemma 3.1.6 that C and L are functorial, and indeed are exact functors of X. Clearly, if $Y \in \mathcal{D}$ then $[Y, LX] = 0$ so $[Y, X] = [Y, CX]$. It follows that C is right adjoint to J.

We can now show that $\mathcal{D}$ is a unital algebraic stable homotopy category. Since $\mathcal{D}$ is localizing, it is certainly a cocomplete triangulated category. By definition, the localizing subcategory generated by $\mathcal{G}$ is $\mathcal{D}$, and the objects of $\mathcal{G}$ are small. It remains to show that $\mathcal{D}$ has a compatible closed symmetric monoidal structure, and that the generators are strongly dualizable. It will then follow from Theorem 1.2.1 that cohomology functors are representable.

We will first show that $\mathcal{D}$ is closed under the smash product in $\mathcal{C}$ (clearly this smash product is compatible with the triangulation and the coproduct). We begin by showing that if $X \in \mathcal{D}$ and $Y \in \mathcal{G}$, then $X \wedge Y \in \mathcal{D}$. This is immediate since the subcategory of such X is localizing and contains $\mathcal{G}$. Similarly, given an arbitrary $X \in \mathcal{D}$, we consider the set of all Y such that $X \wedge Y \in \mathcal{D}$. This is again localizing, and we have just seen that it contains $\mathcal{G}$, so it is all of $\mathcal{D}$ as claimed. Moreover, the unit S lies in $\mathcal{G} \subset \mathcal{D}$.

We still need to construct function objects and show that elements of $\mathcal{G}$ are strongly dualizable. Define $F_{\mathcal{D}}(X, Y)$ to be $CF(X, Y)$. If $X \in \mathcal{D}$ we have

$$[X, F_{\mathcal{D}}(Y, Z)] = [X, F(Y, Z)] = [X \wedge Y, Z].$$

Therefore $F_{\mathcal{D}}(X, Y)$ is adjoint to the smash product on $\mathcal{D}$. Since C is exact, $F_{\mathcal{D}}(-,-)$ is exact as well.

Each object $Z \in \mathcal{G}$ is strongly dualizable in $\mathcal{C}$, so that the map $S \xrightarrow{\eta} F(Z, Z)$ factors through an isomorphism $F(Z, S) \wedge Z \simeq F(Z, Z)$. It follows from our assumptions on $\mathcal{G}$ and the closure of $\mathcal{D}$ under smash products that all the objects just discussed lie in $\mathcal{D}$. Thus

$$F_{\mathcal{D}}(Z, Z) = F(Z, Z) = F(Z, S) \wedge Z = F_{\mathcal{D}}(Z, S) \wedge Z,$$

which means that Z is strongly dualizable in $\mathcal{D}$.

This proves that $\mathcal{D}$ is a unital algebraic stable homotopy category. It is clear that $J\colon \mathcal{D} \longrightarrow \mathcal{C}$ is a geometric morphism, with right adjoint C, and that C preserves the unit.

Consider a family $\{X_i\}$ of objects of $\mathcal{C}$. We then have a cofibration

$$\coprod_i CX_i \rightarrow \coprod_i X_i \rightarrow \coprod_i LX_i.$$

The first term lies in $\mathcal{D}$. Moreover, for any $Z \in \mathcal{G}$ we have

$$[Z, \coprod_i LX_i] = \bigoplus_i [Z, LX_i] = 0.$$

Thus $\coprod_i CX_i$ has the universal property characterizing $C \coprod_i X_i$, which means that C preserves coproducts.

Recall that the category of fractions $\mathcal{C}[\mathcal{S}^{-1}]$ has the same objects as $\mathcal{C}$, and that there is a functor $Q\colon \mathcal{C} \longrightarrow \mathcal{C}[\mathcal{S}^{-1}]$ which sends the maps in $\mathcal{S}$ to isomorphisms. Moreover, Q is the initial example of such a functor: given any functor $F\colon \mathcal{C} \longrightarrow \mathcal{E}$ which inverts $\mathcal{S}$, there is a factorization $F \simeq F'Q$, unique up to natural isomorphism. Clearly $C\colon \mathcal{C} \longrightarrow \mathcal{D}$ inverts $\mathcal{S}$, so $C \simeq C'Q$ for some functor $C'\colon \mathcal{C}[\mathcal{S}^{-1}] \longrightarrow \mathcal{D}$. We also write $J' = QJ\colon \mathcal{D} \longrightarrow \mathcal{C}[\mathcal{S}^{-1}]$. It follows that

$$C'J' \simeq C'QJ \simeq CJ \simeq 1\colon \mathcal{D} \longrightarrow \mathcal{D}.$$

On the other hand, for any object $X = QX \in \mathcal{C}[\mathcal{S}^{-1}]$, we have a map $J'C'QX = CX \xrightarrow{q} X = QX$ which lies in $\mathcal{S}$. Thus q is an isomorphism in $\mathcal{C}[\mathcal{S}^{-1}]$, so $J'C' \simeq 1$. Thus $\mathcal{D} \simeq \mathcal{C}[\mathcal{S}^{-1}]$ as claimed. □

9.2. **Chain complexes.** Most of the non-topological examples of stable homotopy categories we will consider involve chain complexes of objects in additive categories. If an additive category $\mathcal{A}$ is sufficiently nice, the category of chain complexes of objects of $\mathcal{A}$ and chain homotopy classes of maps will satisfy the hypotheses of Theorem 9.1.1. Since all of our algebraic examples fit this description, we follow a somewhat abstract approach, just as we did for triangulated categories.

For this section we assume that $\mathcal{A}$ is an enriched additive category (Definition 1.1.6). If $\mathcal{A}$ happens to be graded, we assume that the closed symmetric monoidal structure has the usual sign conventions. In particular, the symmetric monoidal structure is graded-commutative rather than commutative.

Given any additive category $\mathcal{A}$, we can form the category $\mathrm{Ch}(\mathcal{A})$ of ($\mathbf{Z}$-graded) chain complexes and chain maps. As usual, if X is such a chain complex, we denote its nth component by X_n and its differential by d. It is essentially irrelevant whether the differential raises or lowers degree, but for concreteness we assume it lowers degrees.

Proposition 9.2.1. *If $\mathcal{A}$ is an enriched additive category, then $\mathrm{Ch}(\mathcal{A})$ is an enriched additive category. The obvious inclusion functor $\mathcal{A} \longrightarrow \mathrm{Ch}(\mathcal{A})$ (sending an object M to a complex concentrated in degree zero) is full and faithful, and preserves all structure in sight. If $\mathcal{A}$ is Abelian, then so is $\mathrm{Ch}(\mathcal{A})$.*

Proof. This is all well-known. We can define products and coproducts dimensionwise, so $\mathrm{Ch}(\mathcal{A})$ is complete and cocomplete. If X and Y are chain complexes, we define $(X \wedge Y)_n = \bigoplus_k X_k \otimes Y_{n-k}$. The differential on $X_k \otimes Y_{n-k}$ is $d_X \otimes 1 + (-1)^k \otimes d_Y$. It is easy to check, and standard, that this gives a symmetric monoidal structure on $\mathrm{Ch}(\mathcal{A})$. The component of the twist map $X \wedge Y \longrightarrow Y \wedge X$ sending $X_n \otimes Y_m$ to $Y_m \otimes X_n$ involves a sign $(-1)^{nm}$ as usual.

Similarly, we define $F(X,Y)_n = \prod_k \mathrm{Hom}(X_k, Y_{n+k})$. The component of the differential landing in $\mathrm{Hom}(X_k, Y_{n-1+k})$ is the composite

$$\prod \mathrm{Hom}(X_l, Y_{n+l}) \longrightarrow \mathrm{Hom}(X_{k-1}, Y_{n-1+k}) \oplus \mathrm{Hom}(X_k, Y_{n+k}) \xrightarrow{(-1)^{n+1}\mathrm{Hom}(d_X,1) \oplus \mathrm{Hom}(1,d_Y)} \mathrm{Hom}(X_k, Y_{n-1+k}).$$

It is then straightforward to check that this structure makes $\mathrm{Ch}(\mathcal{A})$ an enriched additive category. If $\mathcal{A}$ is Abelian then we can define kernels and cokernels dimensionwise, and $\mathrm{Ch}(\mathcal{A})$ becomes an Abelian category. □

There is an evident notion of chain homotopy in $\mathrm{Ch}(\mathcal{A})$. For later use, we will describe this somewhat differently than usual. Define I to be the chain complex consisting of R in dimension 1, $R \oplus R$ in dimension 0, and 0 everywhere else. The differential is given by $R \xrightarrow{(1,-1)} R \oplus R$. There are two evident chain maps $R \xrightarrow{i_0, i_1} I$. Two chain maps $f_0, f_1 \colon X \longrightarrow Y$ are then said to be *chain homotopic* if there is a map $H \colon X \wedge I \longrightarrow Y$ such that $H(1 \wedge i_k) = f_k$ for $k = 0, 1$.

It is easy to see that a homotopy $X \wedge I \longrightarrow Y$ is the same thing as a map $X \longrightarrow F(I,Y)$. This is just a formalization of the standard equivalent notions of chain homotopy.

The resulting category of chain complexes and chain homotopy classes of maps will be denoted $\mathcal{K}(\mathcal{A})$.

Proposition 9.2.2. *If $\mathcal{A}$ is an enriched additive category , then $\mathcal{K}(\mathcal{A})$ is an enriched triangulated category. Furthermore, every small (resp., strongly dualizable) object of $\mathcal{A}$ is small (resp., strongly dualizable) in $\mathcal{K}(\mathcal{A})$.*

Proof. The coproduct in $\mathrm{Ch}(\mathcal{A})$ descends to $\mathcal{K}(\mathcal{A})$, as is easy to check. The triangulation on $\mathcal{K}(\mathcal{A})$ is of course well-known. That is, we define the suspension ΣX by $(\Sigma X)_n = X_{n-1}$ with $d_{\Sigma X} = -d_X$. An exact triangle is a sequence isomorphic in $\mathcal{K}(\mathcal{A})$ to one of the form

$$X \xrightarrow{f} Y \xrightarrow{g} Z \xrightarrow{h} \Sigma X$$

where $Z_n = Y_n \oplus X_{n-1}$, $d_Z = \begin{pmatrix} d_Y & f \\ 0 & -d_X \end{pmatrix}$, g is the evident inclusion, and h is the evident projection.

There is also a slightly more flexible way to define triangles. Suppose that $X \xrightarrow{f} Y \xrightarrow{g} Z$ is a sequence of chain complexes, such that in each degree $X_n \xrightarrow{f} Y_n \xrightarrow{g} Z_n$ is a split short exact sequence. (Note that this makes sense even if $\mathcal{A}$ is not Abelian.) This means that we can choose maps $X \xleftarrow{r} Y \xleftarrow{s} Z$ (usually not chain maps) such that

$$rf = 1, \quad rs = 0, \quad gf = 0, \quad gs = 1, \quad sg + fr = 1.$$

It turns out that $h = rd_Y s$ defines a chain map $Z \longrightarrow \Sigma X$, which is independent of the choice of r and s up to homotopy. Moreover, the sequence $X \xrightarrow{f} Y \xrightarrow{g} Z \xrightarrow{h} \Sigma X$ is a triangle. For more details, see [Ive86], for example.

It is straightforward to check that the symmetric monoidal structure defined on $\mathrm{Ch}(\mathcal{A})$ descends to $\mathcal{K}(\mathcal{A})$. Although we have assumed nothing about the exactness of the symmetric monoidal structure on $\mathcal{A}$, cofiber sequences in $\mathcal{K}(\mathcal{A})$ split dimensionwise, so the smash product will automatically be exact on $\mathcal{K}(\mathcal{A})$.

Using the alternative description of a chain homotopy as a map $X \longrightarrow F(I, Y)$, it is easy to check that the function object construction $F(X, Y)$ descends to the homotopy category and is exact there. One must check that the adjointness also descends of course.

It is immediate that strongly dualizable objects of $\mathcal{A}$ give strongly dualizable objects of $\mathcal{K}(\mathcal{A})$, since the relevant function objects and smash product are the same in both categories. To see that a small object M of $\mathcal{A}$ remains small in $\mathcal{K}(\mathcal{A})$, one need only check that $M \wedge I$ is small in $\mathrm{Ch}(\mathcal{A})$. This is easy to do using the standard definition of chain homotopy. □

There is also a natural notion of weak equivalences in $\mathrm{Ch}(\mathcal{A})$ (or $\mathcal{K}(\mathcal{A})$):

Definition 9.2.3. A *quasi-isomorphism* is a chain map $f \colon X \longrightarrow Y$ which induces an isomorphism $H_*X \longrightarrow H_*Y$.

9.3. **The derived category of a ring.** Let R be a commutative ring, and $\mathcal{M}(R)$ the enriched Abelian category of R-modules. We write $\mathcal{K}(R)$ for $\mathcal{K}(\mathcal{M}(R))$. Note that the unit in this category is R, so that

$$\pi_* X = [R, X]_* = H_* X.$$

Theorem 9.3.1. *Let R be a commutative ring. Let $\mathcal{D}(R)$ be the category of fractions obtained from $\mathcal{K}(R)$ by inverting quasi-isomorphisms, so that we have a functor $Q \colon \mathcal{K}(R) \longrightarrow \mathcal{D}(R)$. Then $\mathcal{D}(R)$ can be identified with a subcategory of $\mathcal{K}(R)$, and Q with the right adjoint of the inclusion functor. Moreover:*

(a) *$\mathcal{D}(R)$ is a monogenic stable homotopy theory (with small Hom sets).*
(b) *There is a geometric morphism $J \colon \mathcal{D}(R) \longrightarrow \mathcal{K}(R)$, with right adjoint Q, with the property that $Q(X \otimes JY) = (QX) \otimes Y$.*
(c) *Let X be a bounded-below complex of projectives, and Y an arbitrary complex. Then $[QX, QY] = [X, Y]$.*
(d) *Let X be an arbitrary complex, and Y a bounded-above complex of injectives. Then $[QX, QY] = [X, Y]$.*

(e) *For any short exact sequence* $X \longrightarrow Y \longrightarrow Z$ *of bounded-below complexes in* $\mathrm{Ch}(R)$*, there is a natural exact triangle* $QX \longrightarrow QY \longrightarrow QZ \longrightarrow \Sigma QX$.
(f) *A map* $R \longrightarrow R'$ *induces a stable morphism* $\mathcal{D}(R) \longrightarrow \mathcal{D}(R')$.

We refer to $\mathcal{D}(R)$ as the *derived category* of R. Our approach to the derived category was inspired by [BN93].

Note that in this category $S = R$, so $\pi_* S = R$. Thus, if R is Noetherian, then the results of Section 6 apply. Note also that if R is countable, then $\mathcal{D}(R)$ is a Brown category.

Of course, most of the axioms are well-known for the derived category. See for example [Ver77, Har66, Wei94].

Proof. Parts (a) and (b) are an immediate corollary of Theorem 9.1.1 (with $\mathcal{G} = \{R\}$) and Proposition 9.2.2. The only point to check is that $Q(JX \otimes Y) = (QX) \otimes Y$. This really means that if Y is a cell object then the functor $(-) \otimes Y$ preserves weak equivalences, or equivalently, preserves acyclic complexes. The category of those Y for which this is true is clearly a localizing subcategory, and it clearly contains S; it is therefore the whole of $\mathcal{D}(R)$.

(c): Suppose that X is a bounded-below complex of projectives. Let Z be an exact complex; it is well-known that any chain map $X \longrightarrow Z$ is null-homotopic. The null-homotopy is constructed as usual by induction on the dimension, and we have a place to start because X is bounded below. In particular, if $QX = CX \longrightarrow X \longrightarrow LX$ is the usual cofibration (as in the proof of Theorem 9.1.1), we find that the map $X \longrightarrow LX$ is zero. As L is idempotent, we find that $0 = 1 \colon LX \longrightarrow L^2X = LX$, so that $LX = 0$ and $X = CX = QX$. It follows that $[X, Y] = [QX, QY]$ as claimed.

(d): Suppose that Y is a bounded-above complex of injectives, and X is arbitrary. By the adjoint property of C, we have $[CX, CY] = [CX, Y]$. Any map from an exact complex (such as LX) to Y is null-homotopic, so $[CX, Y] = [X, Y]$.

(e): It is well-known (see [Ive86, Proposition 6.10], for example) that such a sequence can be replaced by a quasi-isomorphic sequence $X' \longrightarrow Y' \longrightarrow Z'$, which is a dimensionwise-split short exact sequence of bounded-below complexes of projectives, and thus a cofiber triangle. Thus, $CX = CX' = X'$ (using (c)). The claim follows.

(f): Given a map $R \longrightarrow R'$, we get a functor $T = R' \otimes_R (-) \colon \mathcal{D}(R) \longrightarrow \mathcal{D}(R')$. This clearly preserves the tensor product and the unit, and sends $\mathcal{G} = \{R\}$ to $\mathcal{G}' = \{R'\}$. In the other direction, we can start with an object of $\mathcal{D}(R')$, regard it as a complex of modules over R, and apply Q; it is not hard to see that this gives a right adjoint to T, so that T is indeed a stable morphism. □

For the interested reader, we describe the closed model structure [Qui67] on the category of chain complexes and chain maps that gives rise to $\mathcal{D}(R)$; this is, essentially, in [Wei94]. We need to identify the weak equivalences, the fibrations, and the cofibrations. A morphism is a weak equivalence if and only if it is a homology isomorphism, a fibration if and only if a dimensionwise surjection, and a cofibration if and only if a dimensionwise injection where the dimensionwise cokernel is cofibrant. A complex is cofibrant if it can be written as an increasing union of complexes so that the associated quotients are complexes of projectives with zero differential.

Notice as well that we can do this same construction if R is a graded ring, using chain complexes of graded R-modules. In that case, we will have two orthogonal directions in which to suspend, so we should consider $\mathcal{D}(R)$ as a multigraded stable homotopy category as in Section 1.3.

We record the following result for use in Section 6.

Let $\mathfrak{p} \leq R$ be a prime ideal, and let $k_\mathfrak{p}$ denote the residue field $(R/\mathfrak{p})_\mathfrak{p}$ of $\mathfrak{p}$. We also write $\overline{k_\mathfrak{p}} = Q(k_\mathfrak{p}) \in \mathcal{D}(R)$.

Proposition 9.3.2. *Suppose that R is Noetherian. Then we have*

$$\operatorname{loc}\langle \overline{k_\mathfrak{p}} \rangle = \operatorname{loc}\langle K(\mathfrak{p}) \rangle = \operatorname{loc}\langle I_\mathfrak{p} \rangle$$

and

$$\langle \overline{k_\mathfrak{p}} \rangle = \langle K(\mathfrak{p}) \rangle = \langle I_\mathfrak{p} \rangle.$$

Proof. First note that, by [Nee92a, Lemma 2.12], we have

$$\langle S \rangle = \coprod \langle \overline{k_\mathfrak{p}} \rangle.$$

The homotopy groups of $\overline{k_\mathfrak{p}}$ are $\mathfrak{p}$-local and $\mathfrak{p}$-torsion. It follows from Theorem 6.1.8 that $\langle \overline{k_\mathfrak{p}} \rangle \leq \langle K(\mathfrak{p}) \rangle$, so that $\overline{k_\mathfrak{p}} \wedge \overline{k_\mathfrak{q}} = 0$ when $\mathfrak{p} \neq \mathfrak{q}$. This means that each $\overline{k_\mathfrak{p}}$ is smash-complemented.

Next, we can use the pairing $QX \otimes QY \longrightarrow Q(X \otimes Y)$ to make $\overline{k_\mathfrak{p}}$ into a ring object, such that $\pi_* \overline{k_\mathfrak{p}} = k_\mathfrak{p}$ is a field. It follows from Propositions 3.7.2 and 3.7.3 that $\overline{k_\mathfrak{p}}$ is a field object, that $\langle \overline{k_\mathfrak{p}} \rangle$ is a minimal Bousfield class, and that $\operatorname{loc}\langle \overline{k_\mathfrak{p}} \rangle$ is a minimal localizing subcategory.

We next claim that $K(\mathfrak{p}) \in \operatorname{loc}\langle \overline{k_\mathfrak{p}} \rangle$. It is easy to reduce to the case in which R is a local ring with maximal ideal $\mathfrak{p}$, so we shall assume this. Let M be a $\mathfrak{p}$-torsion R-module. Write $M_k = \{m \in M \mid \mathfrak{p}^k m = 0\}$, so that $M = \bigcup_k M_k$ and M_k/M_{k-1} is a vector space over $k_\mathfrak{p}$. It follows (using part (e) of Theorem 9.3.1) that $Q(M_k) \in \operatorname{loc}\langle \overline{k_\mathfrak{p}} \rangle$. The map from the sequential colimit of the objects $Q(M_k)$ to QM is a quasi-isomorphism of bounded-below complexes of projectives, hence a homotopy equivalence, which shows that $QM \in \operatorname{loc}\langle \overline{k_\mathfrak{p}} \rangle$. Now suppose that $X \in \mathcal{D}(R)$ is such that $\pi_k(X) = 0$ for $k < 0$ and $\pi_k X$ is $\mathfrak{p}$-torsion for all k. We define $X_0 = X$, and observe that there is a natural map $Q(\pi_0 X_0) \longrightarrow X_0$; we write X_1 for the cofiber. Note that $\pi_k X_1 = 0$ for $k < 1$, and $\pi_* X_1$ is $\mathfrak{p}$-torsion. We define objects X_k and cofibrations $Q(\pi_k X_k) \longrightarrow X_k \longrightarrow X_{k+1}$ in the evident way. We write X^k for the fiber of the map $X \longrightarrow X_k$. It is easy to see that $X^k \in \operatorname{loc}\langle \overline{k_\mathfrak{p}} \rangle$, and that $\pi_m X^k = \pi_m X$ for $k > m+1$. By arguments similar to those of Proposition 2.3.1, we see that X is the sequential colimit of the X^k, and thus X lies in $\operatorname{loc}\langle \overline{k_\mathfrak{p}} \rangle$.

In particular, this shows that $I_\mathfrak{p}$ and $K(\mathfrak{p})$ lie in $\operatorname{loc}\langle \overline{k_\mathfrak{p}} \rangle$ (recall that $\pi_* S = R$ is concentrated in degree zero in this context). As $K(\mathfrak{p}) \neq 0 \neq I_\mathfrak{p}$ and $\operatorname{loc}\langle \overline{k_\mathfrak{p}} \rangle$ is minimal, we see that $\operatorname{loc}\langle \overline{k_\mathfrak{p}} \rangle = \operatorname{loc}\langle K(\mathfrak{p}) \rangle = \operatorname{loc}\langle I_\mathfrak{p} \rangle$ as claimed. It follows that $\langle \overline{k_\mathfrak{p}} \rangle = \langle K(\mathfrak{p}) \rangle = \langle I_\mathfrak{p} \rangle$. □

Example 9.3.3. Let k be a field, and $R = k[\![x,y]\!]/(xy)$. Let $\mathfrak{p} = (x,y)$ be the maximal ideal. Then $K(\mathfrak{p})$ is the finite complex

$$R \xleftarrow{(x,y)} R^2 \xleftarrow{\left(\begin{smallmatrix} y \\ -x \end{smallmatrix}\right)} R \leftarrow 0 \leftarrow \dots$$

and $\overline{k_\mathfrak{p}}$ is the infinite complex

$$R \xleftarrow{(x,y)} R^2 \xleftarrow{\left(\begin{smallmatrix} y & 0 \\ 0 & x \end{smallmatrix}\right)} R^2 \xleftarrow{\left(\begin{smallmatrix} x & 0 \\ 0 & y \end{smallmatrix}\right)} R^2 \xleftarrow{\left(\begin{smallmatrix} y & 0 \\ 0 & x \end{smallmatrix}\right)} R^2 \leftarrow \dots$$

As R is not regular, k has infinite projective dimension as an R-module, and $\overline{k_{\mathfrak{p}}}$ is not a small object.

9.4. **Homotopy categories of equivariant spectra.** In this section, we will recall enough equivariant stable homotopy theory from [LMS86] to conclude that the homotopy category of G-spectra (based on a complete G-universe, where G is a compact Lie group) satisfies our axioms. We will not adopt the most modern approach [EKMM95, Smi] because it is not necessary for the results of this section. We should also point out that the homotopy category of non-equivariant spectra (the case $G = 1$) has been known for a long time to satisfy the axioms for a monogenic stable homotopy category [Vog70, Ada74, Mar83, EKMM95, Smi].

Fix a compact Lie group G.

Definition 9.4.1. Let $\mathcal{U}$ be an inner product space isomorphic to $\mathbf{R}^\infty$, with an orthogonal action of G. $\mathcal{U}$ is a *G-universe* if every finite-dimensional subrepresentation of $\mathcal{U}$ occurs infinitely often in $\mathcal{U}$ and if the trivial representation is a subrepresentation of $\mathcal{U}$. A G-universe is *complete* if every finite-dimensional representation of G is a subrepresentation of $\mathcal{U}$.

If G is trivial, all universes are isomorphic to each other, though non-canonically. However, if G is nontrivial, each isomorphism class of universes will give rise to a distinct stable homotopy category. It turns out to be important to consider the set of isotropy groups of points of $\mathcal{U}$, or equivalently

$$\mathrm{Isotropy}(\mathcal{U}) = \{H \mid G/H \text{ embeds in } \mathcal{U}\}.$$

This is clearly closed under conjugation, and it is also closed under intersection. Indeed, the isotropy subgroup of $(x, y) \in \mathcal{U} \times \mathcal{U}$ is just the intersection of the isotropy subgroups of x and y, and $\mathcal{U} \times \mathcal{U}$ is G-isomorphic to $\mathcal{U}$.

To discuss, even briefly, G-spectra, we need first to recall G-spaces. A G-space is a compactly generated weak Hausdorff space with a continuous action of G. A G-map between two G-spaces is just a G-equivariant continuous map. A based G-space is a G-space with a distinguished point which is fixed by G. A based G-map is a G-map that preserves the basepoints. Given two G-spaces X and Y, $X \times Y$ is the G-space on which G acts diagonally. This allows us to define the smash product $X \wedge Y$ of two based G-spaces as the quotient of $X \times Y$ by the one-point union $X \vee Y$. We can also define $F(X, Y)$, the space of based maps from X to Y, with G acting by conjugation. Given a representation V of G, we have the associated one-point compactification S^V; we consider this as a based G-space, with basepoint at infinity. We denote $S^V \wedge X$ by $\Sigma^V X$, and call this the Vth suspension of X. As usual, suspension has an adjoint $\Omega^V X = F(S^V, X)$.

Given a G-space X and a subgroup H of G, we denote the H-fixed point space by X^H. A weak equivalence of based G-spaces is a based G-map $f\colon X \to Y$ that induces isomorphisms $\pi_*(X^H) \to \pi_*(Y^H)$ of all homotopy groups (relative to the basepoint) of all fixed point sets by closed subgroups H.

Definition 9.4.2. Suppose that $\mathcal{U}$ is a G-universe. A *prespectrum* X is a collection of based G-spaces $X(V)$ for each finite-dimensional subrepresentation V of $\mathcal{U}$, together with G-equivariant maps $\sigma_{V,W}\colon \Sigma^{W-V} X(V) \to X(W)$ for each $V \le W$. Here $W - V$ denotes the orthogonal complement of V in W. The maps $\sigma_{V,W}$ are required to satisfy the transitivity conditions: $\sigma_{V,V}$ is the identity, and if $U \le V \le W$ then $\sigma_{U,W} = \sigma_{V,W} \circ (\Sigma^{W-V} \sigma_{U,V})$. A *spectrum* is a prespectrum such that the adjoints $X(V) \to \Omega^{W-V} X(W)$ of the structure maps are homeomorphisms.

Note that the notions of prespectrum and spectrum depend on both the group G and the universe $\mathcal{U}$. Note as well that there is an evident notion of maps of prespectra and spectra, which are simply maps that commute with the structure maps. This makes the category of spectra $G\mathcal{S}^{\mathcal{U}}$ a full subcategory of the category of prespectra. One of the most important points in this theory is that the inclusion functor from spectra to prespectra has a left adjoint L. This makes the category of spectra have both arbitrary limits and arbitrary colimits, by taking colimits in the category of prespectra and then applying L. One can also use L to define the smash product of a G-space and a spectrum, by smashing space-wise and then applying L.

The category of spectra is enriched over the category of G-spaces, so we can define $X \wedge Y$ when X is a G-space and Y is a spectrum. This gives a natural notion of homotopy in the category of spectra. Indeed, by allowing G to act trivially on the unit interval I, we have cylinder objects $I_+ \wedge X$. We can then define homotopy in the usual way.

For an integer $n \geq 0$, we define the spectrum S^n by applying L to the prespectrum which is $S^V \wedge S^n$ at the representation V, where we are thinking of S^n as a based G-space with trivial G-action, and where the structure maps are isomorphisms. For $n < 0$, we define the spectrum S^n by applying L to the prespectrum which is S^{V-n} for all representations V which contain n, the sum of n copies of the trivial representation, and which is the basepoint otherwise, with the evident structure maps. We then define S^n_H for all closed subgroups $H \leq G$ by $S^n_H = G/H_+ \wedge S^n$. Given a spectrum X, we then define its homotopy group $\pi^n_H(X)$ as the homotopy classes of maps from S^n_H to X. A map $f : X \to Y$ of spectra is then defined to be a weak equivalence if it induces an isomorphism on all homotopy groups. We denote the category of spectra with weak equivalences inverted by $\bar{h}G\mathcal{S}^{\mathcal{U}}$. This is an honest category with small Hom sets, by [LMS86, Section I.6].

The next theorem follows from the first three chapters of [LMS86].

Theorem 9.4.3. *The category $\bar{h}G\mathcal{S}^{\mathcal{U}}$ is an enriched triangulated category (Definition* 1.1.6*). For any closed subgroup H of G, the object $S^0_H = G/H_+ \wedge S$ is small, and the localizing subcategory generated by the S^0_H is all of $\bar{h}G\mathcal{S}^{\mathcal{U}}$. If $H \in \mathrm{Isotropy}(\mathcal{U})$, then $G/H_+ \wedge S$ is strongly dualizable.*

Given the last statement of the theorem, it is natural to define

$$\mathcal{G} = \{G/H_+ \wedge S \mid H \in \mathrm{Isotropy}(\mathcal{U})\}.$$

Proof. This is all contained in the first three chapters of [LMS86], but we recall some of the structure. It is easy to see that the coproduct in spectra descends to the homotopy category. A cofiber sequence is any sequence isomorphic in $\bar{h}G\mathcal{S}^{\mathcal{U}}$ to a sequence

$$X \xrightarrow{f} Y \xrightarrow{g} Z \xrightarrow{h} \Sigma X$$

where $Z = Y \cup_f CX$ is defined as the evident pushout in the category of spectra (as is CX), g is the inclusion, and h is the map obtained by collapsing Y. Then the results of [LMS86, III, Section 2] show that this gives a triangulation on $\mathcal{K}^{\mathcal{U}}_G$, and that fiber sequences, defined analogously using pullbacks and function spaces, are equivalent to cofiber sequences (up to a sign).

The symmetric monoidal structure is a little tricky in [LMS86] because it does not come from a symmetric monoidal structure on the category of spectra. This problem has been fixed more recently [EKMM95, Smi]. The problem is that the smash product of two prespectra indexed on $\mathcal{U}$ is naturally indexed on $\mathcal{U}\times\mathcal{U}$, rather than on $\mathcal{U}$. But one can choose an equivariant linear isometry between $\mathcal{U}\times\mathcal{U}$ and $\mathcal{U}$ and use this, together with L, to make a structure that becomes symmetric monoidal on $\bar{h}G\mathcal{S}^{\mathcal{U}}$. This, and the analogous function spectrum construction, is all contained in [LMS86, II,Section 3].

We learn from [LMS86, Lemma I.5.3] that $G/H_+\wedge S$ is always small. The CW approximation theorem [LMS86, Chapter I] implies that any object in $\bar{h}G\mathcal{S}^{\mathcal{U}}$ is in the localizing subcategory generated by the S^0_H. Moreover, when $H\in \text{Isotropy}(\mathcal{U})$, we see from [LMS86, Theorem III.2.7] that $G/H_+\wedge S$ is strongly dualizable. □

Unfortunately, the category $\bar{h}G\mathcal{S}^{\mathcal{U}}$ is not a stable homotopy category as we have defined it, unless the universe is complete, since otherwise some of the generators S_H need not be strongly dualizable. In fact, Lewis has shown (personal communication) that S_H is never strongly dualizable unless $H\in \text{Isotropy}(\mathcal{U})$. This may be a flaw with our axiom system, or one could interpret it as an unpleasant feature of incomplete universes.

One can always take the "stable hull" of the category $\bar{h}G\mathcal{S}^{\mathcal{U}}$ to obtain a stable homotopy category.

Corollary 9.4.4. *Let $\mathcal{S}^{\mathcal{U}}_G$ be the localizing subcategory generated by the S_H such that $H\in \text{Isotropy}(\mathcal{U})$. Then $\mathcal{S}^{\mathcal{U}}_G$ is a unital algebraic Brown category.*

Proof. First, we claim that

$$G/H_+\wedge G/K_+\wedge S = (G/H\times G/K)_+\wedge S\in \mathcal{S}^{\mathcal{U}}_G$$

whenever H and K are in Isotropy($\mathcal{U}$). From [LMS86], we find that $G/H\times G/K$ is a finite G-CW complex. If there is a cell of type $G/L\times e^n$ in $G/H\times G/K$, then there must be a point with isotropy group L. But then L is an intersection of a conjugate of H with a conjugate of K, so $L\in \text{Isotropy}(\mathcal{U})$. Thus $G/H_+\wedge G/K_+\wedge S$ is a finite G-CW spectrum built from the $G/L_+\wedge S^n$ where $L\in\text{Isotropy}(\mathcal{U})$, and in particular is in $\mathcal{S}^{\mathcal{U}}_G$.

Similarly, the Wirthmüller isomorphism [LMS86, Chapter II] shows that

$$D(G/H_+\wedge S)\in \mathcal{S}^{\mathcal{U}}_G.$$

Thus Theorem 9.1.1 applies, showing that $\mathcal{S}^{\mathcal{U}}_G$ is a unital algebraic stable homotopy category. To see that $\mathcal{S}^{\mathcal{U}}_G$ is actually a Brown category, we use Theorem 4.1.5. The maps between the generators are given by the values of the Burnside ring Mackey functor, which is countable, by the results of [LMS86, Chapter 5], in particular [LMS86, Corollary V.9.4]. □

There is a closed model structure on the category of spectra whose associated homotopy category is $\mathcal{S}^{\mathcal{U}}_G$, at least when $\mathcal{U}$ is complete. The fibrations are spacewise fibrations, and the weak equivalences are spacewise weak equivalences of G-spaces. (A weak equivalence of G-spaces is a weak equivalence on each fixed point set.) The cofibrations are defined by the left lifting property. Hopkins [Hop] has given a proof that this does indeed give a closed model structure.

9.5. **Cochain complexes of B-comodules.** Let B be a commutative Hopf algebra over a field k. Write Comod(B) for the category of left B-comodules, and $\mathcal{K}(B)$ for the homotopy category of chain complexes in Comod(B). As with the derived category of modules over a ring, this needs a little modification before it becomes a stable homotopy category. In this case the right thing to consider is the homotopy category $\mathcal{C}(B)$ of chain complexes of injective comodules.

We work with chain complexes (so that the differential decreases degrees), for consistency with previous sections. Thus, an injective resolution of a comodule will be concentrated in negative degrees. Of course, everything can be translated by the usual prescription $C^i = C_{-i}$.

Recall that a comodule M is *simple* if it is nontrivial, but has no proper nontrivial subcomodules.

Theorem 9.5.1. *Let B be a commutative Hopf algebra over a field k. Then $\mathcal{C}(B)$ is a unital algebraic stable homotopy category, with a geometric morphism $L\colon \mathcal{K}(B) \longrightarrow \mathcal{C}(B)$, whose right adjoint is the inclusion functor. We may take*

$$\mathcal{G} = \{LM \mid M \text{ is a simple comodule }\}.$$

Moreover, L sends complexes of finite total dimension over k to strongly dualizable objects.

If $B = (kG)^*$ where G is a p-group and $p = \operatorname{char}(k)$, then $\mathcal{C}(B)$ is monogenic. The same applies if B is graded and connected (and we consider only graded comodules).

If M and N are comodules then

$$[LM, LN]_* = \operatorname{Ext}^*_B(M, N).$$

A map $B \longrightarrow B'$ of Hopf algebras gives rise to a stable morphism (see Definition 3.4.1) $\mathcal{C}(B) \longrightarrow \mathcal{C}(B')$.

This theorem will be proved after a number of auxiliary results.

Note that in $\mathcal{C}(B)$ we have $L(k) = S$ and thus

$$\pi_* S = [S, S]_* = \operatorname{Ext}^*_B(k, k).$$

If B is finite-dimensional, then Friedlander and Suslin have shown that this is Noetherian [FS]. (This was known in special cases earlier: for $B = (kG)^*$ where G is a finite group [Eve61], for B a finite-dimensional graded connected commutative Hopf algebra [Wil81], and for $B = V(L)^*$ where L is a finite-dimensional restricted Lie algebra [FP87]). If in addition, k (along with its suspensions, in the graded case) is the only simple comodule, then the results of Section 6 apply. For instance, the nilpotence theorem (Corollary 6.1.10) provides a new way to detect nilpotence in $\operatorname{Ext}^*_{kG}(M, M)$ for G a finite p-group and M a finitely generated kG-module.

We begin with some results about the category of B-comodules. It is well-known that the category of modules over a commutative ring is enriched, but it is less well-known that the category of comodules over a commutative Hopf algebra is also enriched.

Note that the dual vector space B^* is a (typically non-commutative) k-algebra.

Definition 9.5.2. A B^*-module M is *tame* if for every $m \in M$, the generated submodule $B^* m$ has finite dimension over k.

Lemma 9.5.3. *Let B be a Hopf algebra over a field k.*

(a) *Given a left B-comodule M and $m \in M$, then the subcomodule generated by m is finite-dimensional over k.*

(b) *There is an isomorphism of categories (which is the identity on objects) between (left) B-comodules and the full subcategory of tame (left) B^*-modules.*

(c) *The resulting inclusion functor J from the category of (left) B-comodules to the category of (left) B^*-modules has a right adjoint R.*
(d) *The category of (left) B-comodules is complete and cocomplete.*
(e) *If B is commutative, then the category of (left) B-comodules is enriched.*

Proof. (a): Choose a basis $\{b_i\}$ for B. Relative to this basis, write $\psi(m) = \sum b_i \otimes m_i$, where all but finitely many of the m_i are 0. Let M' denote the vector space span of the m_i; we claim that M' is a comodule. This is a consequence of coassociativity, and we leave it to the reader.

(b): Given a left B-comodule M, define a B^*-module structure by the composite

$$B^* \otimes M \xrightarrow{1\otimes\psi} B^* \otimes B \otimes M \xrightarrow{\mathrm{ev}\otimes 1} M$$

where $\mathrm{ev}\colon B^* \otimes B \longrightarrow k$ is the evaluation map. One can check that the subcomodule generated by m coincides with the sub B^*-module generated by m, so by part (a) these are all finite-dimensional. This correspondence clearly defines a functor.

We now define the inverse functor. Let N be a B^*-module, with action map $B^* \otimes N \longrightarrow N$. By adjunction we get a map $N \longrightarrow \mathrm{Hom}_k(B^*, N)$. There is an inclusion $B \otimes N \xrightarrow{i} \mathrm{Hom}_k(B^*, N)$ whose adjoint is the evaluation map tensored with N. The image of i is precisely the set of maps f which factor through a finite-dimensional quotient of B^*. In particular, if N is tame then the map $N \longrightarrow \mathrm{Hom}_k(B^*, N)$ will factor through i and give a B-comodule structure on N.

(c): Given an arbitrary B^*-module N, define RN to be the set of all $n \in N$ such that the submodule generated by n is finite-dimensional. Then RN is clearly a submodule, and R is both a left inverse and a right adjoint to the inclusion functor.

(d): Coproducts of comodules can be defined in the usual way, by just taking the direct sum of the underlying vector spaces and giving it the evident comodule structure. To define products, we use R. Given a family of comodules $\{M_i\}$, define $\prod M_i = R(\prod JM_i)$. Then adjointness guarantees that this is a product in the category of B-comodules. Note however that the product of infinitely many nontrivial comodules could easily be trivial. Since the category of B-comodules is Abelian, it has arbitrary (co)limits whenever it has arbitrary (co)products.

(e): The symmetric monoidal structure on the category of B-comodules is well-known. Given B-comodules M and N, we define the comodule structure on $M \otimes_k N$ by the composite

$$M \otimes N \xrightarrow{\psi\otimes\psi} B \otimes M \otimes B \otimes N \xrightarrow{1\otimes T\otimes 1} B \otimes B \otimes M \otimes N \xrightarrow{\mu\otimes 1\otimes 1} B \otimes M \otimes N.$$

To make the category of B-comodules enriched, we also need a notion of function object. It is easier to do this for B^*-modules, so suppose that M and N are B^*-modules such that every principal submodule is finite-dimensional. Dual to the multiplication and conjugation on B we have maps $\Delta^*\colon B^* \longrightarrow \mathrm{Hom}_k(B, B^*)$ and $\chi^*\colon B^* \longrightarrow B^*$. We are going to define a B^*-module structure on $\mathrm{Hom}_k(M, N)$ by a horrendous formula, which is necessary since we are not assuming that B is finite-dimensional. Recall we have chosen a basis $\{b_i\}$ for B; we let $\{b_i^*\}$ denote the dual basis for B^*. Given $f \in \mathrm{Hom}_k(M, N)$ and $u \in B^*$, we define uf by the formula

$$uf(x) = \sum_i b_i^* f[\chi^*((\Delta^* u)(b_i))x].$$

Since M and N are both tame, this sum is in fact finite. Indeed, B^*x is finite-dimensional, so $f(B^*x)$ is as well; thus $b_i^* f(B^*x)$ is zero for almost all i. We leave it to the reader to check that this gives a B^*-module structure on $\operatorname{Hom}_k(M,N)$.

Now, given B-comodules M and N, let us define the function comodule $F(M,N)$ to be $R\operatorname{Hom}_k(JM,JN)$. Then one can verify that the function comodule is right adjoint to the tensor product as required. □

We next investigate the structure of injective comodules.

Lemma 9.5.4.

(a) *The forgetful functor U from comodules to vector spaces has a right adjoint $V \mapsto B \otimes V$, with coaction map $\psi = \Delta \otimes 1$. We refer to $B \otimes V$ as the* cofree *or* extended *comodule on V.*
(b) *For any comodule M we have $B \otimes M \simeq B \otimes UM$. Here the left hand side has the structure described in part* (e) *of Lemma* 9.5.3*, and the left hand side has the comodule structure described in* (a).
(c) *A comodule I is injective if and only if it is a retract of a cofree comodule.*
(d) *If $\{I_\alpha\}$ is a collection of injective comodules, then $\coprod I_\alpha$ and $\prod I_\alpha$ are injective.*
(e) *If I is an injective comodule and M is arbitrary, then $I \otimes M$ and $F(M,I)$ are injective.*

Proof. (a): This is well-known, and easy to check.

(b): The isomorphism is as follows (cf. [Mar83, Proposition 12.4]):

$$\begin{array}{ccc} B \otimes M & \longleftrightarrow & B \otimes UM \\ b \otimes m & \longmapsto & \sum_i bb_i \otimes m_i \\ \sum_i b\chi(b_i) \otimes m_i & \longleftarrow & b \otimes m \end{array}$$

(c): It is clear from the adjunction $\operatorname{Comod}(B)(M, B \otimes V) \simeq \operatorname{Hom}_k(M,V)$ that cofree comodules (and thus their retracts) are injective. Conversely, suppose that I is injective. We have an embedding of comodules $k \longrightarrow B$, and thus an embedding

$$I = k \otimes I \to B \otimes I \simeq B \otimes UI.$$

Because I is injective, this splits, so I is a retract of the cofree comodule $B \otimes UI$.

(d): It is formal (in any category) that products of injectives are injective. For coproducts, use (c).

(e): Let M be an arbitrary comodule. Using (b) and (c), we find that $B \otimes M$ is injective. Using (b) again, we conclude that $I \otimes M$ is injective whenever I is. We also have

$$\operatorname{Comod}(B)(N, F(M,I)) = \operatorname{Comod}(B)(N \otimes M, I).$$

This is clearly an exact functor of N, so $F(M,I)$ is injective. □

Lemma 9.5.5. *Every simple comodule is finite-dimensional, and the collection of isomorphism classes of simple comodules forms a set.*

Proof. Let S be a simple comodule, and $m \in S$ a nonzero element. We know by part (a) of Lemma 9.5.3 that m lies in a finite-dimensional subcomodule of S. This must be all of S, because S is simple. It follows that S is a cyclic B^*-module; there is clearly only a set of these (up to isomorphism). □

Proof of Theorem 9.5.1. The strategy of the proof is very similar to that of Theorem 9.3.1, except we have to localize rather than colocalize. That is, we begin with the category $\mathcal{K}(B)$ of chain complexes of B-comodules and chain homotopy classes of chain maps. By Proposition 9.2.2, $\mathcal{K}(B)$ is an enriched triangulated category. Note as well that any finite-dimensional comodule is small and strongly dualizable in the category of comodules, so will also be in $\mathcal{K}(B)$. Indeed, if M and N are finite-dimensional, so is $\operatorname{Hom}(jM, jN)$, so $F(M,N) = \operatorname{Hom}(jM, jN)$. In particular, since any simple comodule is necessarily finite-dimensional by Lemma 9.5.5, simple comodules are small and strongly dualizable.

But, of course, the simple comodules do not generate $\mathcal{K}(B)$. Ordinarily in this situation we would look at the localizing subcategory generated by the simple comodules, but if we did that in this case we would not get $\operatorname{Ext}_B(k,k)$ as the homotopy of S. So instead we look at $\mathcal{C}(B)$, the full subcategory of $\mathcal{K}(B)$ consisting of (all complexes chain homotopy equivalent to) complexes of injectives. (This is in fact a Bousfield localization.)

Using Lemma 9.5.4, we see that $\mathcal{C}(B)$ is a localizing ideal and a colocalizing coideal in $\mathcal{K}(B)$. In particular, it is closed under products, coproducts, tensor products and function objects. However, it does not contain the unit k.

Let $k \longrightarrow L$ be the cobar resolution of the comodule k. (Any other resolution would do, but we take the cobar resolution to be definite.) Thus L is a complex of injectives, whose homology is k, concentrated in degree zero. The complex L itself is concentrated in degrees less than or equal to zero. We also write C for the fiber of the map $k \longrightarrow L$, so that $H_*C = 0$. This means that C is contractible as a complex of vector spaces, so that $H_*(C \otimes X) = 0$ for any complex X. Finally, we write $CX = C \otimes X$ and $LX = L \otimes X$.

Note L is a functor $\mathcal{K}(B) \longrightarrow \mathcal{C}(B)$. We next show that it is left adjoint to the inclusion $J\colon \mathcal{C}(B) \longrightarrow \mathcal{K}(B)$.

To do so, we first recall the well-known fact that if Y is a bounded-above chain complex (of objects in any Abelian category) with no homology, and U is a complex of injectives, then every map from Y to U is chain homotopic to the zero map. This is proved, as usual, by induction, and since X is bounded above there is a place to start.

Now suppose that $X \in \mathcal{K}(B)$ and $U \in \mathcal{C}(B)$. Then C is bounded above and acyclic, and $F(X,U)$ is a complex of injectives, so

$$[CX, U] = [C, F(X,U)] = 0.$$

It follows from the fibration $C \longrightarrow k \longrightarrow L$ that

$$[LX, U] = [X, U] = [X, JU].$$

In other words, L is left adjoint to J, as claimed.

It follows immediately that $L \simeq 1$ on $\mathcal{C}(B)$, in other words that $L \otimes U \simeq U$ whenever $U \in \mathcal{C}(B)$. Thus L is the unit of the smash product on $\mathcal{C}(B)$, and $\mathcal{C}(B)$ becomes an enriched triangulated category. This also means that L is a geometric morphism.

It follows by juggling adjunctions that we have an internal version of the adjunction: $F(LX, U) \simeq F(X, U)$ whenever U is a complex of injectives.

Now let X be a complex of finite total dimension over k, so that X is strongly dualizable in $\mathcal{K}(B)$. We now show that LX is strongly dualizable in $\mathcal{C}(B)$. The dual of LX in $\mathcal{C}(B)$ is

$$D_{\mathcal{C}(B)}(LX) = F(LX, L) \simeq F(X, L) = F(X, k) \otimes L.$$

More generally, for any $U \in \mathcal{C}(B)$ we have

$$F(LX, U) \simeq F(X, U) = F(X, k) \otimes U.$$

On the other hand, $L \otimes U \simeq U$ so

$$D_{\mathcal{C}(B)}(LX) \otimes U \simeq F(X, k) \otimes L \otimes U \simeq F(X, k) \otimes U \simeq F(LX, U).$$

Thus LX is strongly dualizable as claimed. In particular, this applies to LM when M is a simple comodule.

Now we must show that injective resolutions of simple comodules form a set of weak generators for $\mathcal{C}(B)$. Suppose that

$$X = (\dots \xleftarrow{d_{n-1}} X_{n-1} \xleftarrow{d_n} X_n \xleftarrow{d_{n+1}} \dots)$$

is a complex of injectives, and $[LM, X]_* = [M, X]_* = 0$ for every simple comodule M. Let ZX_n denote the cocycles in dimension n and let BX_n denote the boundaries in dimension n. The exact sequence

$$0 \to ZX_{n+1} \xrightarrow{i} X_{n+1} \xrightarrow{p} BX_n \to 0$$

(where p is just the coboundary d) gives rise to an exact sequence

$$0 \to \operatorname{Hom}_B(M, ZX_{n+1}) \xrightarrow{f} \operatorname{Hom}_B(M, X_{n+1}) \\ \xrightarrow{g} \operatorname{Hom}_B(M, BX_n) \to \operatorname{Ext}^1_B(M, ZX_{n+1}) \to 0$$

since X_n is injective. On the other hand, any map $M \xrightarrow{\alpha} ZX_n$ is a chain map from M (as a complex in degree 0) to X of degree n. Thus, since $[M, X]_* = 0$, it must be null-homotopic. That is, there must be a lift of α to $\beta\colon M \to X_{n-1}$. The composite $\operatorname{Hom}_B(M, X_{n+1}) \xrightarrow{g} \operatorname{Hom}_B(M, BX_n) \xrightarrow{h} \operatorname{Hom}_B(M, ZX_n)$ is therefore surjective. Since h is monic, it follows that h is an isomorphism and g is surjective. In particular, $\operatorname{Ext}^1_B(M, ZX_{n+1}) = 0$ for every simple comodule M.

Lemma 9.5.7 then shows that ZX_{n-1} is injective. There are then splittings $r\colon X_{n-1} \to ZX_{n-1}$ of i and $q\colon BX_n \to X_{n-1}$ of p. In particular, BX_n is also injective. By considering the exact sequence

$$0 \to BX_n \to ZX_n \to H_n X \to 0$$

we find an exact sequence

$$0 \to \operatorname{Hom}_B(M, BX_n) \xrightarrow{h} \operatorname{Hom}_B(M, ZX_n) \to \operatorname{Hom}_B(M, H_n X) \to 0.$$

Since h is an isomorphism for any simple comodule M, we find that

$$\operatorname{Hom}_B(M, H_n X) = 0$$

for every such M. By Lemma 9.5.6, $H_n X = 0$, so $ZX_n = BX_n$.

We can now define a chain homotopy $D\colon X_n \to X_{n-1}$ by the composite $X_n \xrightarrow{r} ZX_n \simeq BX_n \xrightarrow{q} X_{n-1}$. It is then easy to see that $dD + Dd$ is the identity of X, so X is a contractible chain complex.

It follows that $\mathcal{C}(B)$ is a unital algebraic stable homotopy category, with

$$\mathcal{G} = \{LM \mid M \text{ is a simple comodule }\}.$$

If k is the only simple comodule, then $\mathcal{C}(B)$ is monogenic. It is well known that this is the case when G is a p-group with $p = \text{char}(k)$, and $B = (kG)^*$. Similarly if B is graded and connected.

Let M and N be comodules. Then LN is a complex of injectives with homology $H_*(L \otimes N) = H_*(L) \otimes N = N$, in other words an injective resolution of N. Moreover, we know that $[LM, LN]_* = [M, LN]_*$. It follows that

$$[LM, LN]_* = \text{Ext}_B^*(M, N)$$

as claimed.

Finally, suppose we have a map $f: B \longrightarrow B'$ of Hopf algebras. Given a complex of injective B-comodules, we can think of it as a complex of B'-comodules through f. We then apply L to get a complex of injective B'-comodules. This gives a functor $\mathcal{C}(B) \longrightarrow \mathcal{C}(B')$, which we claim is a stable morphism; we leave the details to the reader. □

We still owe the reader Lemmas 9.5.6 and 9.5.7.

Lemma 9.5.6. *The set $\{M\}$ of simple comodules weakly generates* Comod(B). *In other words, for every nonzero comodule N, there is an inclusion $M \hookrightarrow N$ of a simple comodule into N.*

Proof. By Lemma 9.5.3, every B-comodule $N \neq 0$ has a finite-dimensional subcomodule $N' \neq 0$. Clearly every finite-dimensional B-comodule $N' \neq 0$ has a simple sub-comodule (by induction on dimension). □

Lemma 9.5.7. *Suppose that J is a comodule such that $\text{Ext}_B^1(M, J) = 0$ for all simple comodules M. Then J is injective.*

Proof. We first note that $\text{Ext}_B^1(F, J) = 0$ for all finite-dimensional (over k) comodules F. Indeed, we prove this by induction on the dimension. Any comodule of dimension 1 is simple. Given a nonzero finite-dimensional comodule F, there is a nonzero simple comodule $M \leq F$. By considering the short exact sequence

$$0 \longrightarrow M \longrightarrow F \longrightarrow F/M \longrightarrow 0$$

we find an exact sequence

$$\text{Ext}_B^1(F/M, J) \longrightarrow \text{Ext}_B^1(F, J) \longrightarrow \text{Ext}_B^1(M, J)$$

and so, by induction, $\text{Ext}_B^1(F, J) = 0$.

Now, suppose that we have an arbitrary inclusion of comodules $M \leq N$, and a map $f: M \longrightarrow J$. We must extend f to N. Consider the set of pairs (P, g) where $M \leq P \leq N$ and $g: P \longrightarrow J$ is an extension of f, ordered in the evident way. By applying Zorn's lemma, we find a maximal such extension (N', g'). We claim that $N' = N$. Indeed, suppose not. Then choose an element $n \in N$ but not in N'. Let N'' denote the subcomodule generated by N' and n. Then N''/N' is generated by n, and hence, by Lemma 9.5.3, N''/N is finite-dimensional. Thus, $\text{Ext}_B^1(N''/N', J) = 0$, and so the map $\text{Hom}_B(N'', J) \longrightarrow \text{Hom}_B(N', J)$ is onto. Thus there is an extension of g' to N'', violating the fact that (N', g') is maximal. Hence $N' = N$, as required. □

Note that the homotopy groups in $\mathcal{C}(B)$ of a complex of injectives X are

$$\pi_* X = [L, X]_* = H_*(PX),$$

where P denotes the primitive functor. If k is the only simple comodule, as will occur for example when B is graded connected, then $H_*(PX) = 0 \Leftrightarrow X = 0$.

As usual, $\mathcal{C}(B)$ is the homotopy category of a closed model structure on the Abelian category of chain complexes of comodules. We will describe the structure without giving proofs. The cofibrations are dimensionwise inclusions, and the fibrations are dimensionwise surjections with kernel a complex of injectives. The weak equivalences are generalized homotopy isomorphisms—that is, a map $f\colon X \longrightarrow Y$ is a weak equivalence if and only if it induces isomorphisms $[Z, L \wedge X] \longrightarrow [Z, L \wedge Y]$ for all finite-dimensional comodules Z.

Remark 9.5.8. Suppose that A is a finite-dimensional graded connected cocommutative Hopf algebra over a field k; suppose also that A is a Koszul algebra [Pri70, BGS] with Koszul dual $A^!$. Then by [BGS, Theorem 16] there is an equivalence of triangulated categories between $\mathcal{F}_{\mathcal{C}(A^*)}$ and $\mathcal{F}_{\mathcal{D}(A^!)}$. Therefore we have a classification of the thick subcategories of these, by Example 6.1.4. Note that this agrees with the classification of thick subcategories of $\mathcal{F}_{\mathcal{C}((kG)^*)}$, for G a p-group.

9.6. **The stable category of B-modules.** In this section, we let B be a finite-dimensional commutative Hopf algebra over a field k, and we study comodules over B. It would be equivalent to consider modules over a finite-dimensional cocommutative Hopf algebra, in view of the following result.

Proposition 9.6.1. *There is an isomorphism (which is the identity on objects) between the categories of B-comodules and B^*-modules.*

Proof. Any B^*-comodule is clearly tame (Definition 9.5.2), given that B^* has finite dimension. The claim therefore follows from part (b) of Lemma 9.5.3. □

We can now define the stable categories which we wish to study.

Definition 9.6.2. Given two comodules M and N over B, write

$$\mathrm{Hom}_B(M,N)_0 = \{f \in \mathrm{Hom}_B(M,N) \mid f \text{ factors through an injective comodule }\}.$$

This is a subspace of $\mathrm{Hom}_B(M,N)$, so we can define

$$\underline{\mathrm{Hom}}_B(M,N) = \mathrm{Hom}_B(M,N)/\mathrm{Hom}_B(M,N)_0.$$

The composition map $\mathrm{Hom}_B(L,M) \otimes \mathrm{Hom}_B(M,N) \longrightarrow \mathrm{Hom}_B(L,N)$ descends to give a well-defined composition

$$\underline{\mathrm{Hom}}_B(L,M) \otimes \underline{\mathrm{Hom}}_B(M,N) \longrightarrow \underline{\mathrm{Hom}}_B(L,N).$$

We can therefore define a category $\mathrm{StComod}(B)$ whose objects are B-comodules, and whose morphisms are the sets $\underline{\mathrm{Hom}}_B(M,N)$.

In this section we prove the following theorems, and we also make a few remarks about classifying thick subcategories and detecting nilpotence.

Theorem 9.6.3. *Suppose that B is a finite-dimensional commutative Hopf algebra over a field k; then* $\mathrm{StComod}(B)$ *is a unital algebraic stable homotopy category. If k is countable then it is a Brown category. If k is the only simple B-comodule, then* $\mathrm{StComod}(B)$ *is monogenic.*

It turns out that $\mathrm{StComod}(B)$ is equivalent to a Bousfield localization (in fact, a finite localization—Definition 3.3.4) of $\mathcal{C}(B)$. Because of this, there is a strong relation between the two categories; see Lemma 3.5.6 and Proposition 9.6.8, for example.

Theorem 9.6.4. *Let B be a finite-dimensional commutative Hopf algebra over a field k. Let $L_B^f\colon \mathcal{C}(B) \to \mathcal{C}(B)$ denote finite localization with respect to the thick subcategory generated by B (as a complex of injectives concentrated in degree 0). Then we have an equivalence of stable homotopy categories*

$$L_B^f \mathcal{C}(B) \simeq \mathrm{StComod}(B).$$

Before proving Theorems 9.6.3 and 9.6.4, we need three lemmas.

Lemma 9.6.5. *There is an isomorphism $B \simeq \mathrm{Hom}(B, k)$ of B-comodules. A comodule is projective if and only if it is injective.*

Proof. See [LS69, p. 85]. □

Lemma 9.6.6. *A chain complex X in $\mathcal{C}(B)$ is L_B^f-local if and only if $H_*(X) = 0$.*

Proof. X is L_B^f-local if and only if $[B, X]_* = 0$. But $H_*(X) = \pi_*(B \wedge X) = [DB, X]_* = [B, X]_*$, since B is self-dual. □

Recall that L denotes the cobar resolution of the ground field k (which has finite dimension in each degree). Given any comodule M, we have both an injective resolution

$$M \to L \otimes M$$

and a projective resolution

$$\mathrm{Hom}(L, k) \otimes M = \mathrm{Hom}(L, M) \to M.$$

We can then splice these resolutions together to form the *Tate complex* $t_B(M)$:

$$\ldots \to \mathrm{Hom}(L_{-1}, M) \to \mathrm{Hom}(L_0, M) \to L_0 \otimes M \to L_{-1} \otimes M \to \ldots.$$

Here $L_0 \otimes M$ is in degree zero, so $\mathrm{Hom}(L_{-k}, M)$ is in degree $k+1$.

Because projectives and injectives are the same, $t_B(M)$ is an element of $\mathcal{C}(B)$. Since projective and injective resolutions of M are unique up to chain equivalence, $t_B(M)$ is independent of our choice of L. Since $t_B(M)$ clearly has no homology, it is L_B^f-local by the preceding lemma.

Lemma 9.6.7. *For any comodule M, $t_B(M)$ is the L_B^f-localization of LM.*

Proof. There is certainly a map $LM \xrightarrow{f} t_B(M)$, and we have already seen that $t_B(M)$ is L_B^f-local. The cofiber of f is the projective resolution $\mathrm{Hom}(L, M)$ of M, and it suffices to show that $\mathrm{Hom}(L, M)$ is in the localizing subcategory $\mathcal{D} = \mathrm{loc}\langle B \rangle$ generated by B. We prove more generally that any bounded below complex of injectives lies in $\mathcal{D}$.

First consider an injective comodule J thought of as a complex concentrated in a single degree. Then J is a retract of a coproduct of copies of B, so it lies in $\mathcal{D}$.

Next, let I be a bounded-below complex of injectives. Let $I(k)$ be the truncated complex

$$\ldots \to 0 \to I_k \to I_{k-1} \to \ldots.$$

Note that $I(k) = 0$ for $k \ll 0$, and there are cofibrations $I(k) \to I(k+1) \to I(k+1)/I(k)$. By the previous paragraph, $I(k+1)/I(k) \in \mathcal{D}$, so $I(k) \in \mathcal{D}$ for all k. Let $I(\infty)$ be the sequential colimit of the $I(k)$'s, so that we have a map $I(\infty) \to I$ with cofiber I' say. It is not hard to see that I' is a bounded below complex of injectives with $H_*I' = 0$, so I' is contractible and $I(\infty) \simeq I$. It follows that $I \in \mathcal{D}$ as claimed. □

We now prove Theorems 9.6.3 and 9.6.4 simultaneously.

Proof of Theorems 9.6.3 *and* 9.6.4. The strategy of the proof is first to define structures on $\mathrm{StComod}(B)$, then to construct an equivalence of categories between $L_B^f \mathcal{C}(B)$ and $\mathrm{StComod}(B)$ that preserves these structures. Since L_B^f is smashing, as finite localizations always are, this shows that with these structures $\mathrm{StComod}(B)$ is a stable homotopy category. We leave it to the reader to check that the structures we define on $\mathrm{StComod}(B)$ are equivalent to similar structures on $\mathrm{StMod}(B^*)$.

We begin by defining the standard structures on $\mathrm{StComod}(B)$ that will make it into a stable homotopy category. First, the coproduct of stable comodules is the same as the coproduct of comodules. The suspension of a stable comodule M is defined by embedding M into the injective module $B \otimes M$ and taking the cokernel. A cofiber sequence in $\mathrm{StComod}(B)$ is any sequence isomorphic to

$$M \xrightarrow{f} N \xrightarrow{g} P \xrightarrow{h} \Sigma M$$

where

$$0 \to M \xrightarrow{f} N \xrightarrow{g} P \to 0$$

is a short exact sequence of comodules, and where h is any map such that the following diagram commutes:

$$\begin{array}{ccccccccc}
0 & \longrightarrow & M & \longrightarrow & B \otimes M & \longrightarrow & \Sigma M & \longrightarrow & 0 \\
 & & {\scriptstyle =}\big\downarrow & & \big\downarrow & & {\scriptstyle -h}\big\downarrow & & \\
0 & \longrightarrow & M & \xrightarrow{f} & N & \xrightarrow{g} & P & \longrightarrow & 0.
\end{array}$$

The smash product of two stable comodules is just the tensor product $M \otimes N$ and the function object is just $\mathrm{Hom}_k(M, N)$. (Since B is finite, the subtleties in Lemma 9.5.3 do not arise.) We leave it to the reader to check that these do define functors in the stable category and that they remain adjoint.

We still have to prove that with these structures and with generating set consisting of the simple comodules, $\mathrm{StComod}(B)$ is a stable homotopy category. Rather than proving this directly, we will construct an equivalence of categories between $\mathrm{StComod}(B)$ and $L_B^f \mathcal{C}(B)$ that preserves coproducts, cofiber sequences, smash products, and function objects. Since L_B^f is a finite localization, it is in particular smashing, so $L_B^f \mathcal{C}(B)$ is a unital algebraic stable homotopy category, and the result follows.

We begin by considering the functor $G\colon \mathrm{Comod}(B) \to \mathcal{C}(B)$ that takes M to $L \otimes M$. This functor preserves coproducts and takes the tensor product to the smash product. Also, if

$$0 \to M \xrightarrow{f} N \xrightarrow{g} P \to 0$$

is a short exact sequence, then the cofiber of $G(f)$ is a complex of injectives beginning in degree 0 whose only homology group is P, so it is equivalent to $G(P)$.

The composite functor $L_B^f G$ is equivalent to t_B by Lemma 9.6.7. Since any injective module lies in $\mathrm{loc}\langle B \rangle$, the functor t_B factors through the stable category to give a functor

$$t_B\colon \mathrm{StComod}(B) \to L_B^f\mathcal{C}(B).$$

Since L_B^f preserves the coproduct and the smash product, so does t_B. By considering the short exact sequence

$$0 \to M \xrightarrow{f} B \otimes M \to \Sigma M \to 0$$

we find that the cofiber of $t_B(f)$ is $t_B(\Sigma M)$. But $t_B(B \otimes M)$ is trivial, so this cofiber is $\Sigma t_B(M)$. Thus t_B preserves the suspension, and it is then easy to see that it preserves cofiber sequences as well.

We now construct an inverse to t_B. Consider an object $X = (\dots \to X_1 \to X_0 \to X_{-1} \to \dots)$ which is L_B^f-local, i.e., which is an acyclic chain complex of injectives. Define

$$u(X) = \ker(X_0 \to X_{-1}) = \mathrm{image}(X_1 \to X_0).$$

This is clearly functorial for chain maps. Moreover, if $f\colon X \to Y$ is null-homotopic then $u(f)$ factors through the injective comodule Y_1. It follows that u gives a functor $L_B^f\mathcal{C}(B) \to \mathrm{StComod}(B)$. It is easy to check that t_B and u define an equivalence of categories. It follows by adjointness that t_B preserves function objects, so in fact t_B is an equivalence of stable homotopy categories. □

The proof of Theorem 9.6.4 makes it clear that the homotopy of S in the stable module category of kG is just the Tate cohomology of G, hence the notation t_B. For a general commutative Hopf algebra B, the category of complexes of injectives with no homology will not in general be colocalizing, so there is no way one could perform Bousfield localization and land in it. On the other hand, one can always perform Bousfield localization L_H with respect to the ordinary homology functor H on $\mathcal{C}(B)$. Denote the fiber of the localization map $X \to L_H X$ by $C_H X$. Then one could define the Tate cohomology of a general Hopf algebra B to be $[C_H S, C_H S]$. This agrees with the usual Tate cohomology of a finite group. Mislin [Mis] has recently given an extension of Tate cohomology to arbitrary groups. We have not checked whether his definition agrees with ours.

Theorem 9.6.4 implies that there is a strong tie between the two categories $\mathcal{C}(B)$ and $\mathrm{StComod}(B)$. Here is one example, motivated by Rickard's classification [Ric] of thick subcategories of small objects in $\mathrm{StMod}(kG)$, for G a p-group.

Proposition 9.6.8. *There is a one-to-one correspondence between the thick subcategories of $\mathcal{F}_{\mathrm{StComod}(B)}$ and the nonzero thick subcategories of $\mathcal{F}_{\mathcal{C}(B)}$.*

Before proving this proposition, we need a lemma.

Lemma 9.6.9. *A comodule M is small in* $\mathrm{StComod}(B)$ *if and only if M is isomorphic in* $\mathrm{StComod}(B)$ *to a finite-dimensional comodule.*

Proof. First suppose M is a finite-dimensional comodule. Then $t_B(M) = L_B^f LM$ is small in $L_B^f\mathcal{C}(B)$, since LM is small in $\mathcal{C}(B)$. Thus M is small in $\mathrm{StComod}(B)$. Conversely, any comodule M which is small in $\mathrm{StComod}(B)$ is in the thick subcategory generated by the simple comodules, since $\mathrm{StComod}(B)$ is a unital algebraic stable homotopy category. Any simple comodule is finite-dimensional, and the property of being isomorphic to a finite-dimensional comodule is preserved under suspensions, cofibrations, and retracts. The only one of these claims that is not immediately clear is the closure under retracts. To see this, we use [Mar83, Proposition 13.13] (and the comments immediately following it) to write any comodule M uniquely as $I \oplus M'$, where I is injective and M' has no injective summands. Of course, M and M' are isomorphic in StComod(B), and conversely, if M and N are isomorphic in StComod(B), then M' and N' are isomorphic as comodules [Mar83, Proposition 14.1]. Thus, if N is a retract of M in StComod(B), then N' is a retract of M' as comodules. In particular, if M is isomorphic to a finite-dimensional comodule, then M' must be finite-dimensional, and so N' is as well. □

Proof of Proposition 9.6.8. Let $L = L_B^f$, so that StComod(B) is equivalent to $\mathcal{C}(B)_L$. We claim that there is a correspondence

$$\begin{array}{ccc} \left\{ \begin{array}{c} \text{nonzero thick subcats} \\ \text{in } \mathcal{F}_{\mathcal{C}(B)} \end{array} \right\} & \longleftrightarrow & \left\{ \begin{array}{c} \text{thick subcats} \\ \text{in } \mathcal{F}_{\mathcal{C}(B)_L} \end{array} \right\} \\ \mathcal{D} & \longmapsto & \text{thick}\langle L\mathcal{D}\rangle \\ L^{-1}\mathcal{D}' \cap \mathcal{F}_{\mathcal{C}(B)} & \longleftarrow & \mathcal{D}' \end{array}$$

To see that these maps give a one-to-one correspondence, note first that the preceding lemma shows that $L_B^f\colon \mathcal{F}_{\mathcal{C}(B)} \to \mathcal{F}_{\mathrm{StComod}(B)}$ is surjective on objects. From this, it is straightforward to check that, if $\mathcal{D}'$ is a thick subcategory of $\mathcal{F}_{\mathrm{StComod}(B)}$, then $\text{thick}\langle L(L^{-1}\mathcal{D}' \cap \mathcal{F}_{\mathcal{C}(B)})\rangle = \mathcal{D}'$.

Conversely, suppose that $\mathcal{D}$ is a thick subcategory of $\mathcal{F}_{\mathcal{C}(B)}$, and let $\overline{\mathcal{D}} = L^{-1}L\mathcal{D}$ be the L-replete thick subcategory generated by $\mathcal{D}$. By Lemma 3.5.6 it suffices to show that $\overline{\mathcal{D}} \cap \mathcal{F}_{\mathcal{C}(B)} = \mathcal{D}$.

For all $X \in \mathcal{D}$, we have $L_{\mathcal{D}}^f X = 0$. Note that if $Y \xrightarrow{f} Z$ is an L-equivalence, then it is also an $L_{\mathcal{D}}^f$-equivalence; it follows that the full category of all objects X satisfying $L_{\mathcal{D}}^f X = 0$ is thick and L-replete, so it contains $\overline{\mathcal{D}}$. By Theorem 3.3.3, then, if X is finite and in $\overline{\mathcal{D}}$, then $X \in \mathcal{D}$. □

We finish by pointing out a familiar result which is an application of the fact that StMod(kG) is a stable homotopy category.

Theorem 9.6.10 ([QV72, Car81]). *Fix a finite group G, a field k of characteristic $p > 0$, and a finitely-generated kG-module M. Then an element $z \in \mathrm{Ext}_{kG}^*(M, M)$ is nilpotent if and only if $\mathrm{res}_{G,E}(z) \in \mathrm{Ext}_{kE}^*(M, M)$ is nilpotent for every elementary Abelian p-subgroup E of G.*

Proof. Chouinard's theorem [Cho76] says that a kG-module M is projective (i.e., zero in StMod(kG)) if and only if $M{\downarrow}_E$ is a projective kE-module for all elementary Abelian p-subgroups E of G (here, $M{\downarrow}_E$ denotes M restricted to kE). Now, one can easily show that $M{\downarrow}_E$ is a projective kE-module if and only if $k[G/E]^* \otimes M$ is a projective kG-module. In other words, we have the following equality of Bousfield classes:

$$\langle S\rangle = \coprod_{\substack{E \leq G \\ \text{elem.ab.}}} \langle k[G/E]^*\rangle.$$

Hence Theorem 5.1.2 applies. On the other hand, these modules $k[G/E]^*$ represent the homology functors

$$M \mapsto \mathrm{Ext}_{kE}^*(k, M).$$

This finishes the proof. □

Remark 9.6.11.

(a) One can also give a proof of Chouinard's theorem in the language of stable homotopy theory. Of course one needs to use Serre's characterization of elementary Abelian p-groups via products of Bocksteins [Ser65]; given that, however, it is straightforward.
(b) We have shown that Theorem 9.6.10 is a formal consequence of Chouinard's theorem. This is perhaps different from the usual point of view on this result.
(c) A similar result (with the same proof) holds in the category of chain complexes of projective kG-modules—one can imitate the arguments in Section 9.5 to show that this is a stable homotopy category. We leave the details to the reader. If G is a p-group, then this category is monogenic; so since $\mathrm{Ext}^*_{kG}(k,k)$ is Noetherian, then the results of Section 6 apply, giving a different nilpotence theorem in this setting.
(d) One can also use the same arguments to recover the nilpotence theorem for a finite-dimensional cocommutative Hopf algebra B^*—see [Wil81, Palb]. Again, one can work either in $\mathrm{StMod}(B^*)$ or $\mathcal{C}(B_*)$.

10. Future directions

We close the paper by briefly discussing some topics we feel merit further study.

One such topic is the Adams spectral sequence. Given a ring object E in the ordinary stable homotopy category, we can construct a spectral sequence that attempts to calculate $[X,Y]$ in terms of E-homology information [Ada74]. One needs some hypotheses on E to determine the E_2-term of this spectral sequence, and some hypotheses on E, X, and Y to guarantee convergence. This construction of the Adams spectral sequence will certainly work in a monogenic Brown category, but one might hope to be able to construct it more generally. Convergence will certainly be delicate, as it is even with spectra. One might hope for a construction in an algebraic stable homotopy category, and for some convergence results analogous to those in [Bou79b], but we have not investigated this question. It is certainly an important one. It also seems likely that many familiar spectral sequences (such as the Cartan-Eilenberg spectral sequence associated to an extension of Hopf algebras) can be presented in these terms.

10.1. **Grading systems on stable homotopy categories.** There are times when we would like to allow a stable homotopy category to have a more general grading than the $\mathbf{Z}$-grading enjoyed by any triangulated category, or even than the multigrading discussed in Section 1.3. In this section we briefly discuss such grading systems. The basic examples are the possibility of grading over the Picard group (Definition A.2.7), and, in the setting of G-equivariant stable homotopy theory, grading so that $[X,Y]_*$ is a $\mathbf{Z}$-graded Mackey functor. The main advantage of this more complicated grading is that one can make any unital algebraic stable homotopy category monogenic, if one is willing to modify the grading of the category.

Suppose that we have a stable homotopy category $\mathcal{C}$. Consider $(\mathbf{Z},+)$ as a symmetric monoidal category with the only maps being the identity map 1_m and $(-1)_m$ on each object m, with the evident composition. The symmetric monoidal structure is just addition on the objects and multiplication on the maps. It is strictly associative and unital, but not strictly commutative. The commutativity natural transformation is $(-1)^{mn}$ on $m \wedge n$.

Recall that a strict symmetric monoidal functor $F: \mathcal{C} \longrightarrow \mathcal{D}$ between symmetric monoidal categories is a functor equipped with natural isomorphisms $FX \wedge FY \simeq F(X \wedge Y)$ and $FS \simeq S$, which are compatible in the evident sense with the commutativity, associativity and unity maps in $\mathcal{C}$ and $\mathcal{D}$. We shall simply refer to such a thing as a monoidal functor.

If $\mathcal{C}$ is an enriched triangulated category, then we can define a monoidal functor $(\mathbf{Z}, +) \xrightarrow{S} \mathcal{C}$, which takes m to S^m.

Definition 10.1.1. Let $\mathcal{C}$ be an enriched triangulated category.

(a) A *pointed symmetric monoidal category* is a symmetric monoidal category $\mathcal{C}$ equipped with a commutative monoidal functor $(\mathbf{Z}, +) \longrightarrow \mathcal{C}$. There is an evident notion of a pointed functor of pointed symmetric monoidal categories.

(b) A *grading system* on an closed symmetric monoidal triangulated category $\mathcal{C}$ is a small pointed symmetric monoidal category $\mathcal{I}$ and a pointed commutative monoidal functor $G: \mathcal{I} \longrightarrow \mathcal{C}$ such that Ga is strongly dualizable for all $a \in \mathcal{I}$.

(c) Given a grading system G on an closed symmetric monoidal triangulated category, we write S^a for Ga, and we write $\Sigma^a X$ for $S^a \wedge X$. Furthermore, we define $[X, Y]_a = [\Sigma^a X, Y]$. If H is a homology functor, we define $H_a X = H(DS^a \wedge X)$ and if H is a cohomology functor we define $H^a X = H(DS^a \wedge X)$.

(d) Given a grading system G on an closed symmetric monoidal triangulated category, we modify the definitions of thick, localizing, and colocalizing subcategories so as to require them to be closed under smashing with all the S^a, which we think of as a generalized suspension.

(e) An $\mathcal{I}$*-graded stable homotopy category* is a category $\mathcal{C}$ equipped with
   (i) Arbitrary coproducts
   (ii) A triangulation
   (iii) A closed symmetric monoidal structure compatible with the triangulation
   (iv) A grading system $G: \mathcal{I} \longrightarrow \mathcal{C}$ such that the localizing subcategory (in the above sense) generated by S is all of $\mathcal{C}$.

   such that all cohomology functors are representable.

Example 10.1.2. (a) Every closed symmetric monoidal triangulated category admits a grading system where $\mathcal{I} = \mathbf{Z}$.

(b) If $\mathcal{C}$ is a unital algebraic stable homotopy category, then there is always a choice of grading system on $\mathcal{C}$ so that $\mathcal{C}$ becomes an $\mathcal{I}$-graded stable homotopy category. Indeed, we can take for $\mathcal{I}$ a small equivalent subcategory of $\mathcal{F}$ and for G the inclusion functor.

(c) If the Picard category (see Definition A.2.7) of an closed symmetric monoidal triangulated category $\mathcal{C}$ is small, then the inclusion functor Pic $\longrightarrow \mathcal{C}$ defines a grading system on $\mathcal{C}$. Grading over the Picard *group* is not always possible, but can be done if and only if one can choose a monoidal section of the map from the Picard category to the Picard group. In particular, one can always grade over free Abelian subgroups of the Picard group. A bigraded category is an example of this, as is the $RO(G)$ grading on the homotopy category of G-spectra defined over a complete universe [LMS86]. Note, however, that the category of G-spectra is not an $RO(G)$-graded stable homotopy category, because the localizing subcategory generated by the "representation spheres" is not usually the whole category.

(e) Let G be a finite group. The homotopy category of G-spectra (based on a complete universe) admits a grading system over $(\mathbf{Z},+)\times\mathcal{L}$ where $\mathcal{L}$ is the Lindner category of finite G-sets. Here objects are finite G-sets, but maps are arbitrary maps of G-sets together with transfers. The symmetric monoidal structure is given by the product. In this case $[X,Y]_*$, as a contravariant functor from $(\mathbf{Z},+)\times\mathcal{L}$ to Abelian groups, is a graded Mackey functor, and the category of G-spectra becomes a $(\mathbf{Z},+)\times\mathcal{L}$-graded stable homotopy category. See [LMS86] for details.

(f) In any $\mathcal{I}$-graded stable homotopy category, there is a symmetric monoidal structure on the category of contravariant functors from $\mathcal{I}$ to Ab so that π_*S becomes a commutative ring object. In the equivariant case, one recovers the notion of Green functors.

The approach outlined here is a useful one in equivariant stable homotopy theory, but we do not know if it is useful in other stable homotopy categories.

10.2. **Other examples.** To conclude the paper, we provide a partial list of cohomology theories that we think are associated with stable homotopy categories. We have not considered these, for reasons of space, time, and energy.

(a) Ext of modules over a commutative S-algebra in the sense of [EKMM95], possibly in a G-equivariant setting. These certainly form a stable homotopy category, but we have not studied any examples in depth.
(b) Ext of differential graded modules over a commutative differential graded algebra. This example should be very similar to the derived category of an ordinary ring. It is likely that the Adams spectral sequence in this category is the spectral sequence of Eilenberg-Moore that goes from ordinary Ext to differential Ext.
(c) Ext of sheaves of $\mathcal{O}_X$-modules, where X is a scheme.
(d) Continuous cohomology of a topological group, especially of a profinite group.
(e) Ext of comodules over a commutative Hopf algebroid, such as BP_*BP.
(f) Ext of Mackey functors which are modules over a Green functor. This example might help one to understand what to expect for a thick subcategory theorem in equivariant stable homotopy theory.
(g) Motivic cohomology in any of its various guises.

We have also considered generalizing the definition of a stable homotopy category to allow for a non-commutative smash product. This would cover the derived category of bimodules over a non-commutative ring, for example. One might hope to be able to put the following things in this context:

(a) Cyclic cohomology.
(b) Hochschild cohomology.
(c) Ext of bimodules over a noncommutative S-algebra.

It is possible that Hochschild cohomology should be thought of in the same way as Tate cohomology of spectra [GM95]. That is, given an algebra A over a commutative ring B, we should construct the Hochschild cohomology of A as a ring object in the derived category of B-modules. We do not yet understand the relationship between this approach and the idea of making a stable homotopy category of bimodules.

Appendix A. Background from category theory

A.1. **Triangulated categories.** We recall the notion of a triangulated category. We give the definition from [Mar83]; this definition is equivalent to the more usual one as in [Ver77], and see [Nee92b] for yet another set of axioms.

Definition A.1.1. A *triangulation* on an additive category $\mathcal{C}$ is an additive (*suspension*) functor $\Sigma\colon \mathcal{C} \to \mathcal{C}$ giving an automorphism of $\mathcal{C}$, together with a collection Δ of diagrams, called *exact triangles* or *cofiber sequences*, of the form

$$X \to Y \to Z \to \Sigma X$$

such that

1. Any diagram isomorphic to a diagram in Δ is in Δ.
2. Any diagram of the following form is in Δ:

$$0 \to X \xrightarrow{1} X \to 0$$

3. If the first of the following diagrams is in Δ, then so is the second:

$$X \xrightarrow{f} Y \xrightarrow{g} Z \xrightarrow{h} \Sigma X$$

$$Y \xrightarrow{g} Z \xrightarrow{h} \Sigma X \xrightarrow{-\Sigma f} \Sigma Y.$$

4. For any map $f\colon X \to Y$, there is a diagram of the following form in Δ.

$$X \xrightarrow{f} Y \to Z \to \Sigma X$$

5. Suppose we have a diagram as shown below (with h missing), in which the rows lie in Δ and the rectangles commute. Then there exists a (nonunique) map h making the whole diagram commutative.

$$\begin{array}{ccccccc} U & \longrightarrow & V & \longrightarrow & W & \longrightarrow & \Sigma U \\ \downarrow{\scriptstyle f} & & \downarrow & & \downarrow{\scriptstyle h} & & \downarrow{\scriptstyle \Sigma f} \\ X & \longrightarrow & Y & \longrightarrow & Z & \longrightarrow & \Sigma X \end{array}$$

6. Verdier's octahedral axiom holds: Suppose we have maps $X \xrightarrow{v} Y \xrightarrow{u} Z$, and cofiber triangles (X, Y, U), (X, Z, V) and (Y, Z, W) as shown in the diagram. (A circled arrow $U \circ\!\!\!\to X$ means a map $U \to \Sigma X$.) Then there exist maps r and s as shown, making (U, V, W) into a cofiber triangle, such that the following commutativities hold:

$$au = rd \qquad es = (\Sigma v)b \qquad sa = f \qquad br = c$$

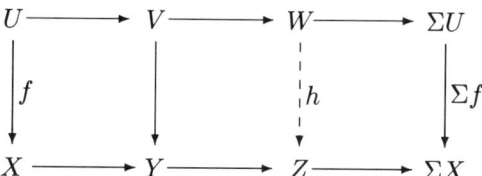

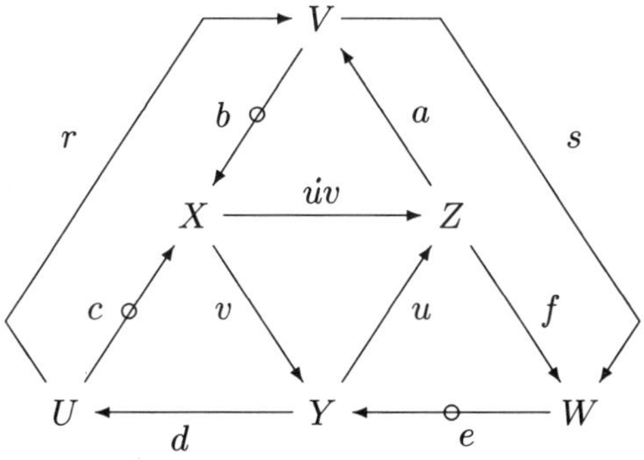

(If u and v are inclusions of CW spectra, this essentially just says that $(Z/X)/(Y/X) = Z/Y$. The diagram can be turned into an octahedron by lifting the outer vertices and drawing an extra line from W to U.)

A category equipped with a triangulation is called a *triangulated category*. Given an exact sequence $\Sigma^{-1}Z \to X \xrightarrow{f} Y \to Z$, we say that Z is the *cofiber* and $\Sigma^{-1}Z$ the *fiber* of f. The cofiber of f is only determined up to unnatural isomorphism in the category of objects under Y and over ΣX. If Z is the cofiber of $f\colon X \to Y$, we will often abuse notation by referring to the sequence $X \xrightarrow{f} Y \to Z$ as a cofiber sequence.

An *exact functor* between triangulated categories is a functor L which is equipped with an equivalence $L\Sigma \simeq \Sigma L$, and which preserves cofiber sequences. More precisely, suppose that $X \to Y \to Z \to \Sigma X$ is a cofiber sequence. We can apply L and use the given equivalence $L\Sigma X = \Sigma LX$ to get a sequence $LX \to LY \to LZ \to \Sigma LX$. The requirement is that this should again be a cofiber sequence.

A natural transformation of exact functors is required to commute with the given suspension equivalences in the obvious sense.

We will use a number of well-known properties of triangulated categories without proof, and usually without explicit mention; see [Mar83] and [Ver77] for more complete references. In particular, in any triangulated category, coproducts and products, when they exist, preserve cofiber sequences.

We will state the (perhaps poorly named) 3×3 lemma here, however. A proof can be found in [BBD82] and an interesting discussion of related matters can be found in [Nee92b].

Lemma A.1.2 (3 × 3 lemma). *Let $\mathcal{C}$ be a triangulated category. Consider a commutative square as shown, in which the rows and columns are cofiber sequences.*

$$\begin{array}{ccccccc}
 & & \Sigma^{-1}X' & & \Sigma^{-1}Y' & & \\
 & & \downarrow & & \downarrow & & \\
\Sigma^{-1}Z'' & \longrightarrow & X'' & \longrightarrow & Y'' & \longrightarrow & Z'' \\
 & & \downarrow & & \downarrow & & \\
\Sigma^{-1}Z & \longrightarrow & X & \longrightarrow & Y & \longrightarrow & Z \\
 & & \downarrow & & \downarrow & & \\
 & & X' & & Y' & &
\end{array}$$

Then there exists an object Z' and maps $Y' \to Z' \leftarrow Z$, such that the following diagram commutes (except that the top left square anticommutes) and the rows and columns are exact.

$$\begin{array}{ccccccc}
\Sigma^{-2}Z' & \longrightarrow & \Sigma^{-1}X' & \longrightarrow & \Sigma^{-1}Y' & \longrightarrow & \Sigma^{-1}Z' \\
\downarrow & & \downarrow & & \downarrow & & \downarrow \\
\Sigma^{-1}Z'' & \longrightarrow & X'' & \longrightarrow & Y'' & \longrightarrow & Z'' \\
\downarrow & & \downarrow & & \downarrow & & \downarrow \\
\Sigma^{-1}Z & \longrightarrow & X & \longrightarrow & Y & \longrightarrow & Z \\
\downarrow & & \downarrow & & \downarrow & & \downarrow \\
\Sigma^{-1}Z' & \longrightarrow & X' & \longrightarrow & Y' & \longrightarrow & Z'
\end{array}$$

It is understood that the map $X' \to Y'$ is the suspension of the map $\Sigma^{-1}X' \to \Sigma^{-1}Y'$, and so on.

A.2. **Closed symmetric monoidal categories.**

Definition A.2.1. A *closed symmetric monoidal category* is a category $\mathcal{C}$ equipped with:

1. A unit object S.
2. A functor $(X, Y) \mapsto X \wedge Y$ from $\mathcal{C} \times \mathcal{C}$ to $\mathcal{C}$, which is associative and commutative up to coherent natural isomorphism, such that $S \wedge X = X$ up to coherent natural isomorphism. We shall call this functor the *smash product*, by analogy with the category of spectra.
3. Function objects $F(X, Y)$, which are functorial contravariantly in X and covariantly in Y, such that $[X, F(Y, Z)] \simeq [X \wedge Y, Z]$, naturally in all three variables.

We shall say that this structure is *compatible* with a given triangulation on $\mathcal{C}$ if:

1. The smash product preserves suspensions. That is, there is a natural equivalence $e_{X,Y} : \Sigma X \wedge Y \to \Sigma(X \wedge Y)$. Furthermore, if we let r_X denote the unital equivalence $X \wedge S \to X$, then we have $\Sigma r_X \circ e_{X,S} = r_{\Sigma X}$, and if $a_{X,Y,Z}$ denotes the associativity isomorphism $(X \wedge Y) \wedge Z \to X \wedge (Y \wedge Z)$, then the

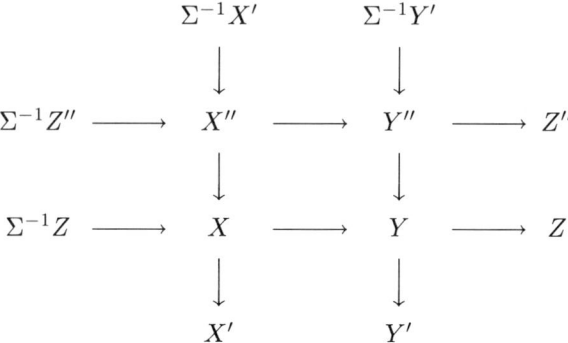

following diagram commutes:

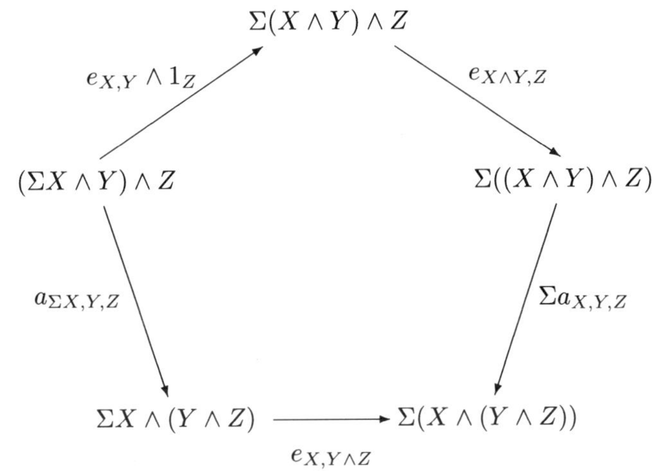

One can easily construct from e isomorphisms $F(\Sigma X, Y) \simeq \Sigma^{-1} F(X,Y)$ and $F(X, \Sigma Y) \simeq \Sigma F(X,Y)$.

2. The smash product is exact. More precisely, suppose that $X \xrightarrow{f} Y \xrightarrow{g} Z \xrightarrow{h} \Sigma X$ is an exact triangle, and that W is an object of $\mathcal{C}$. If we use $e_{X,W}$ to identify $(\Sigma X) \wedge W$ with $\Sigma(X \wedge W)$, then the following triangle is required to be exact:

$$X \wedge W \xrightarrow{f \wedge 1} Y \wedge W \xrightarrow{g \wedge 1} Z \wedge W \xrightarrow{h \wedge 1} \Sigma(X \wedge W)$$

3. The functor $F(X,Y)$ is exact in the second variable in a similar sense, and is exact in the first variable up to sign. That is, suppose $X \xrightarrow{f} Y \xrightarrow{g} Z \xrightarrow{h} \Sigma X$ is an exact triangle, and that W is an object of $\mathcal{C}$. If we use the adjoint of e to identify $F(\Sigma X, W)$ with $\Sigma^{-1} F(X,W)$, then the following triangle is required to be exact:

$$\Sigma^{-1} F(X,W) \xrightarrow{-F(h,1)} F(Z,W) \xrightarrow{F(g,1)} F(Y,W) \xrightarrow{F(f,1)} F(X,W)$$

4. The smash product interacts with the suspension in a graded-commutative manner. That is, the following diagram is commutative for all integers r and s, where T is the twist map (i.e., the commutativity equivalence for the smash product), $S^r = \Sigma^r S$, and the horizontal equivalences come from the equivalence e above together with the symmetric monoidal structure.

$$\begin{array}{ccc} S^r \wedge S^s & \xrightarrow{\simeq} & S^{r+s} \\ {\scriptstyle T}\downarrow & & \downarrow{\scriptstyle (-1)^{rs}} \\ S^s \wedge S^r & \xrightarrow{\simeq} & S^{r+s} \end{array}$$

We really need only require that the last diagram commute with $r = s = 1$; it then follows that each transposition in the symmetric group Σ_n acts as -1 on $S^n = S^1 \wedge \ldots \wedge S^1$, so every permutation acts as its signature. One can deduce from this that the diagram commutes for all r and s.

There are a number of theorems which say approximately the following: if diagrams of a certain type commute in the category of (possibly infinite-dimensional) vector spaces over **C**, then they commute in any closed symmetric monoidal category. See for example [Sol95].

One way to interpret the sign that prevents $F(X, Y)$ from being exact in the first variable is that mapping out of a cofiber sequence should produce a fiber sequence, not a cofiber sequence, and in the stable case, cofiber sequences differ from fiber sequences only by a sign. This sign can usually be ignored, since most of our arguments do not rely on the features of the maps in an exact triangle, but only on the existence of the exact triangle.

Remark A.2.2. Note that, in a closed symmetric monoidal category $\mathcal{C}$, the smash product is always compatible with coproducts. That is, given a family $\{X_i\}$ of objects of $\mathcal{C}$ such that $\coprod X_i$ exists, and given another object Y of $\mathcal{C}$, the coproduct $\coprod(X_i \wedge Y)$ exists, and the natural map $\coprod(X_i \wedge Y) \to (\coprod X_i) \wedge Y$ is an isomorphism. Indeed, we have, for any object Z of $\mathcal{C}$,

$$[(\coprod X_i) \wedge Y, Z] = [\coprod X_i, F(Y, Z)] = \prod [X_i, F(Y, Z)] = \prod [X_i \wedge Y, Z],$$

as required.

Proposition A.2.3. *Suppose that $\mathcal{C}$ is a closed symmetric monoidal category.*

(a) *There is an associative composition map*

$$F(X, Y) \wedge F(Y, Z) \xrightarrow{\circ} F(X, Z).$$

For all X, there is a map $S \xrightarrow{\eta} F(X, X)$ which is a two-sided unit for the composition. This makes $\mathcal{C}$ a category enriched over itself, as in [Kel82].

(b) *Both the smash product and the function object functor are (canonically) enriched functors, and they are adjoint as enriched functors. That is, there is a natural isomorphism*

$$F(X, F(Y, Z)) \simeq F(X \wedge Y, Z).$$

(c) *Coproducts and products in $\mathcal{C}$ are enriched coproducts and products as well. That is, we have equivalences*

$$F(\coprod X_\alpha, Y) \to \prod F(X_\alpha, Y)$$

and

$$F(X, \prod Y_\alpha) \to \prod F(X, Y_\alpha).$$

Proof. The unit map $S \to F(X, X)$ is adjoint to the unit equivalence $S \wedge X \to X$. Because of the adjunction, we have evaluation maps

$$F(X, Y) \wedge X \xrightarrow{\text{ev}} Y$$

and we use these to define the composition map as the adjoint of the composite

$$F(X, Y) \wedge X \wedge F(Y, Z) \xrightarrow{\text{ev} \wedge 1} Y \wedge F(Y, Z) \xrightarrow{\text{ev}} Z.$$

We leave it to the reader to check that this composition is associative and unital. To say that the smash product is an enriched functor means that we have a natural map

$$F(X, X') \wedge F(Y, Y') \to F(X \wedge Y, X' \wedge Y')$$

compatible with the composition and the unit. To construct this map, we take the adjoint to the map

$$F(X, X') \wedge X \wedge F(Y, Y') \wedge Y \xrightarrow{\mathrm{ev} \wedge \mathrm{ev}} X' \wedge Y'.$$

We leave it to the reader to check that this is compatible with composition and the unit and to construct analogous maps for the function object functor.

One way to see that $F(X \wedge Y, Z)$ is naturally equivalent to $F(X, F(Y, Z))$ is to show that they represent the same functor. That is, we have

$$[W, F(X \wedge Y, Z)] = [W \wedge X \wedge Y, Z] = [W \wedge X, F(Y, Z)] = [W, F(X, F(Y, Z))].$$

One can use a similar method to show that coproducts and products behave as expected. □

Recall from [LMS86, Chapter III] the definition of strongly dualizable objects in a closed symmetric monoidal triangulated category: Z is strongly dualizable if the natural map

$$F(Z, S) \wedge X \to F(Z, X)$$

is an isomorphism for all X. We can now see that this natural map is nothing more than composition, if we interpret X as $F(S, X)$. This motivates the following definition.

Definition A.2.4. In a closed symmetric monoidal category $\mathcal{C}$, we denote $F(X, S)$ by DX and refer to it as the *Spanier-Whitehead dual* of X. Note that D is an exact contravariant functor that takes coproducts to products.

Strongly dualizable objects were studied in [LMS86, Chapter III], and are the basis for Spanier-Whitehead duality. We recall the results of [LMS86] in the following theorem.

Theorem A.2.5. *Let $\mathcal{C}$ be a category with a triangulation and a closed symmetric monoidal structure compatible with the triangulation.*

(a) *The full subcategory of strongly dualizable objects is thick and closed under smash products and function objects. In particular, if X is strongly dualizable, so is DX.*

(b) *If X is strongly dualizable, the natural map $X \longrightarrow D^2X$ adjoint to the evaluation map*

$$X \wedge DX \to S$$

is an isomorphism.

(c) *If Y is strongly dualizable and X and Z are arbitrary objects of $\mathcal{C}$, there is a natural isomorphism*

$$F(X \wedge Y, Z) \to F(X, DY \wedge Z).$$

(d) *The natural map*

$$F(X, Y) \wedge F(X', Y') \to F(X \wedge X', Y \wedge Y')$$

is an equivalence when X and X' are strongly dualizable, and also when X is strongly dualizable and $Y = S$. In particular, if X is strongly dualizable, the map

$$DX \wedge DY \to D(X \wedge Y)$$

is an isomorphism for all Y.

(e) *If X or Z is strongly dualizable, the natural map*

$$F(X,Y) \wedge Z \to F(X, Y \wedge Z)$$

is an isomorphism.

(f) *If X is strongly dualizable and $\{Y_\alpha\}$ is a family of objects, then the natural map*

$$X \wedge \prod Y_\alpha \to \prod (X \wedge Y_\alpha)$$

is an isomorphism.

With the exception of the fact that strongly dualizable objects form a thick subcategory, this theorem holds in an arbitrary closed symmetric monoidal category.

Proof. This is all proved in [LMS86] except for two things: the fact that strongly dualizable objects form a thick subcategory, and part (f). The former follows easily from the exactness of $F(-,Y)$. For the latter, we have the equivalences

$$X \wedge \prod Y_\alpha \simeq F(DX, \prod Y_\alpha) \simeq \prod F(DX, Y_\alpha) \simeq \prod (X \wedge Y_\alpha),$$

completing the proof. □

Another useful lemma proved in [LMS86] (see Proposition III.1.3 and its proof) is the following.

Lemma A.2.6. *Suppose that X is a strongly dualizable object in a closed symmetric monoidal category. Then X is a retract of $X \wedge DX \wedge X$.* □

Of course, S is always strongly dualizable in any closed symmetric monoidal category. The Picard category, first introduced into stable homotopy theory by Hopkins [HMS94, Str92], provides another source of strongly dualizable objects.

Definition A.2.7. Let $\mathcal{C}$ be a closed symmetric monoidal category. We say that an object $X \in \mathcal{C}$ is *invertible* if there is an object Z and an isomorphism $X \wedge Z \to S$. Define the *Picard category* to be the full subcategory of invertible objects. We refer to the isomorphism classes of this category with the operation $\wedge$ as the *Picard group*, though in general it may be a proper class rather than a set. We will denote the Picard group by Pic.

Proposition A.2.8. *Let $\mathcal{C}$ be a closed symmetric monoidal category. Then any object X of the Picard category is strongly dualizable. Furthermore, the inverse of X is DX.*

Proof. Let Y denote an inverse of X. Then smashing with Y is an equivalence of $\mathcal{C}$ with itself. Therefore, for all Z, we have

$$[Z, DX] \simeq [Z \wedge X, S] \simeq [Z \wedge X \wedge Y, Y] \simeq [Z, Y].$$

Hence $Y = DX$. Now for all W and Z we also have

$$[W, F(X,Z)] \simeq [W \wedge X, Z] \simeq [W, Z \wedge Y] \simeq [W, DX \wedge Z],$$

Thus $F(X,Z) \simeq DX \wedge Z$ and so X is strongly dualizable. □

REFERENCES

[Ada71] J. F. Adams, *A variant of E. H. Brown's representability theorem*, Topology **10** (1971), 185–198.

[Ada74] J. F. Adams, *Stable homotopy and generalised homology*, The University of Chicago Press, 1974.

[BBD82] A. A. Beilinson, J. N. Bernstein, and P. Deligne, *Faisceaux pervers*, Astérisque **100** (1982).

[BGS] A. A. Beilinson, V. Ginsburg, and W. Soergel, *Koszul duality patterns in representation theory*, J. Amer. Math. Soc., to appear.

[BCR] D. J. Benson, J. F. Carlson, and J. Rickard, *Complexity and varieties for infinitely generated modules, II*, Math. Proc. Camb. Phil. Soc., to appear.

[BL95] J. Block and A. Lazarev, *Homotopy theory and generalized duality for spectral sheaves*, preprint, 1995.

[BN93] M. Bokstedt and A. Neeman, *Colimits in triangulated categories*, Compositio Math. **86** (1993), 209–234.

[Bou79a] A. K. Bousfield, *The Boolean algebra of spectra*, Comment. Math. Helv. **54** (1979), 368–377.

[Bou79b] A. K. Bousfield, *The localization of spectra with respect to homology*, Topology **18** (1979), 257–281.

[Bou83] A. K. Bousfield, *Correction to 'The Boolean algebra of spectra'*, Comment. Math. Helv. **58** (1983), 599–600.

[Car81] J. F. Carlson, *Complexity and Krull dimension*, Representations of Algebras (M. Auslander and E. Lluis, eds.), 1981, Lecture Notes in Mathematics, vol. 903, Springer-Verlag, pp. 62–67.

[Cho76] L. G. Chouinard, *Projectivity and relative projectivity over group rings*, J. Pure Appl. Algebra **7** (1976), 287–302.

[Chr] J. D. Christensen, *Phantom maps and the structure of the stable homotopy category*, Ph. D. thesis in progress.

[DHS88] E. S. Devinatz, M. J. Hopkins, and J. H. Smith, *Nilpotence and stable homotopy theory*, Ann. of Math. (2) **128** (1988), 207–241.

[DS95] W. G. Dwyer and J. Spalinski, *Homotopy theories and model categories*, Handbook of Algebraic Topology (Ioan M. James, ed.), Elsevier, Amsterdam, 1995.

[EKMM95] A. D. Elmendorf, I. Kriz, M. A. Mandell, and J. P. May, *Rings, modules, and algebras in stable homotopy theory*, Amer. Math. Soc. Surveys and Monographs, to appear.

[Eve61] L. Evens, *The cohomology ring of a finite group*, Trans. Amer. Math. Soc. **101** (1961), 224–239.

[FP87] E. M. Friedlander and B. J. Parshall, *Geometry of p-unipotent lie algebras*, J. Algebra **109** (1987), 25–45.

[FS] E. M. Friedlander and A. Suslin, *Cohomology of finite group schemes over a field*, preprint.

[Gol86] J. S. Golan, *Torsion theories*, Pitman monographs and surveys in pure and applied mathematics, vol. 29, Longman Scientific & Technical, 1986.

[GM95] J. P. C. Greenlees and J. P. May, *Generalized Tate cohomology*, Mem. Amer. Math. Soc., vol. 543, American Mathematical Society, 1995.

[Har66] R. Hartshorne, *Residues and duality*, Lecture Notes in Mathematics, vol. 20, Springer-Verlag, 1966.

[Hop] M. J. Hopkins, lectures at M. I. T.

[Hop87] M. J. Hopkins, *Global methods in homotopy theory*, Proceedings of the Durham Symposium on Homotopy Theory (J. D. S. Jones and E. Rees, eds.), 1987, LMS Lecture Note Series 117, pp. 73–96.

[HMS94] M. J. Hopkins, M. E. Mahowald, and H. Sadofsky, *Constructions of elements in Picard groups*, Topology and Representation Theory (E. Friedlander and M. E. Mahowald, eds.), Contemp. Math., vol. 158, 1994, pp. 89–126.

[HS] M. J. Hopkins and J. H. Smith, *Nilpotence and stable homotopy theory II*, preprint.

[HSS] M. Hovey, H. Sadofsky, and N. P. Strickland, *Morava K-theories and localization*, preprint.

[Ive86] B. Iverson, *Cohomology of sheaves*, Universitext, Springer-Verlag, 1986.

[Jen72] C. U. Jensen, *Les foncteurs dérivés de $\varprojlim$ et leurs applications en théorie des modules*, Lecture Notes in Mathematics, vol. 254, Springer-Verlag, 1972.

[Kel82] G. M. Kelly, *Basic concepts of enriched category theory*, London Mathematical Society Lecture Notes, vol. 64, Cambridge University Press, 1982.

[KM95] I. Kriz and J. P. May, *Operads, algebras, modules, and motives*, Astérisque **233** (1995).

[LS69] R. G. Larson and M. E. Sweedler, *An associative orthogonal bilinear form for Hopf algebras*, Amer. J. Math. **91** (1969), 75–94.

[LMS86] L. G. Lewis, Jr., J. P. May, and M. Steinberger, *Equivariant stable homotopy theory*, Lecture Notes in Mathematics, vol. 1213, Springer-Verlag, 1986.

[Mar83] H. R. Margolis, *Spectra and the Steenrod algebra*, North-Holland, 1983.

[Mat89] H. Matsumura, *Commutative ring theory*, Cambridge studies in advanced mathematics, vol. 8, Cambridge University Press, 1989.

[Mil92] H. R. Miller, *Finite localizations*, Boletin de la Sociedad Matematica Mexicana **37** (1992), 383–390, This is a special volume in memory of José Adem, in book form. The editor is Enrique Ramírez de Arellano.

[Mis] G. Mislin, *Tate cohomology for arbitrary groups via satellites*, Topology and its Applications **56** (1994), 293–300.

[Nee92a] A. Neeman, *The chromatic tower for D(R)*, Topology **31** (1992), 519–532.

[Nee92b] A. Neeman, *Some new axioms for triangulated categories*, J. Algebra **139** (1992), 221–255.

[Nee95] A. Neeman, *On a theorem of Brown and Adams*, preprint, 1995.

[Pal92] J. H. Palmieri, *Self-maps of modules over the Steenrod algebra*, J. Pure Appl. Algebra **79** (1992), 281–291.

[Pala] J. H. Palmieri, *Nilpotence for modules over the mod* 2 *Steenrod algebra I*, Duke Math. J., to appear.

[Palb] J. H. Palmieri, *A note on the cohomology of finite dimensional cocommutative Hopf algebras*, preprint.

[Pri70] S. B. Priddy, *Koszul resolutions*, Trans. Amer. Math. Soc. **152** (1970), 39–60.

[Qui67] D. G. Quillen, *Homotopical algebra*, Lecture Notes in Mathematics, vol. 43, Springer-Verlag, 1967.

[QV72] D. G. Quillen and B. B. Venkov, *Cohomology of finite groups and elementary abelian subgroups*, Topology **11** (1972), 317–318.

[Rav84] D. C. Ravenel, *Localization with respect to certain periodic homology theories*, Amer. J. Math. **106 (1)** (1984), 351–414.

[Rav92] D. C. Ravenel, *Nilpotence and periodicity in stable homotopy theory*, Annals of Mathematics Studies, vol. 128, Princeton University Press, 1992.

[Ric] J. Rickard, personal communication.

[Sch] S. Schwede, *Spectra in model categories and applications to the algebraic cotangent complex*, J. Pure Appl. Algebra, to appear.

[Ser65] J.-P. Serre, *Sur la dimension cohomologique des groupes profinis*, Topology **3** (1965), 413–420.

[Smi] J. H. Smith, in preparation.

[Sol95] S. Soloviev, *Proof of a S. MacLane conjecture*, Category Theory and Computer Science (David Pitt, David E. Rydeheard, and Peter Johnstone, eds.), 1995, Springer Lecture Notes in Computer Science 953, pp. 59–80.

[Str92] N. P. Strickland, *On the p-adic interpolation of stable homotopy groups*, Adams Memorial Symposium on Algebraic Topology Volume 2 (N. Ray and G. Walker, eds.), 1992, London Mathematical Society Lecture Notes 176, pp. 45–54.

[Str] N. P. Strickland, *No small objects*, preprint.

[Tho] R. W. Thomason, *The classification of triangulated subcategories*, Compositio Math., to appear (1995).

[Ver77] J.-L. Verdier, *Catégories dérivées*, Cohomologie Etale (SGA $4\frac{1}{2}$) (P. Deligne, ed.), 1977, Lecture Notes in Mathematics, vol. 569, Springer-Verlag, pp. 262–311.

[Vog70] R. Vogt, *Boardman's stable homotopy category*, Aarhus Univ. Lect. Notes, vol. 21, Aarhus University, 1970.

[Wei94] C. A. Weibel, *An introduction to homological algebra*, Cambridge studies in advanced mathematics, vol. 38, Cambridge University Press, 1994.

[Wil81] C. Wilkerson, *The cohomology algebras of finite dimensional Hopf algebras*, Trans. Amer. Math. Soc. **264** (1981), 137–150.

Index

Editorial Information

To be published in the *Memoirs*, a paper must be correct, new, nontrivial, and significant. Further, it must be well written and of interest to a substantial number of mathematicians. Piecemeal results, such as an inconclusive step toward an unproved major theorem or a minor variation on a known result, are in general not acceptable for publication. *Transactions* Editors shall solicit and encourage publication of worthy papers. Papers appearing in *Memoirs* are generally longer than those appearing in *Transactions* with which it shares an editorial committee.

As of March 31, 1997, the backlog for this journal was approximately 8 volumes. This estimate is the result of dividing the number of manuscripts for this journal in the Providence office that have not yet gone to the printer on the above date by the average number of monographs per volume over the previous twelve months, reduced by the number of issues published in four months (the time necessary for preparing an issue for the printer). (There are 6 volumes per year, each containing at least 4 numbers.)

A Copyright Transfer Agreement is required before a paper will be published in this journal. By submitting a paper to this journal, authors certify that the manuscript has not been submitted to nor is it under consideration for publication by another journal, conference proceedings, or similar publication.

Information for Authors and Editors

Memoirs are printed by photo-offset from camera copy fully prepared by the author. This means that the finished book will look exactly like the copy submitted.

The paper must contain a *descriptive title* and an *abstract* that summarizes the article in language suitable for workers in the general field (algebra, analysis, etc.). The *descriptive title* should be short, but informative; useless or vague phrases such as "some remarks about" or "concerning" should be avoided. The *abstract* should be at least one complete sentence, and at most 300 words. Included with the footnotes to the paper, there should be the 1991 *Mathematics Subject Classification* representing the primary and secondary subjects of the article. This may be followed by a list of *key words and phrases* describing the subject matter of the article and taken from it. A list of the numbers may be found in the annual index of *Mathematical Reviews*, published with the December issue starting in 1990, as well as from the electronic service e-MATH [**telnet e-MATH.ams.org** (or **telnet 130.44.1.100**). Login and password are **e-math**]. For journal abbreviations used in bibliographies, see the list of serials in the latest *Mathematical Reviews* annual index. When the manuscript is submitted, authors should supply the editor with electronic addresses if available. These will be printed after the postal address at the end of each article.

Electronically prepared papers. The AMS encourages submission of electronically prepared papers in AMS-TeX or AMS-LaTeX. The Society has prepared author packages for each AMS publication. Author packages include instructions for preparing electronic papers, the *AMS Author Handbook*, samples, and a style file that generates the particular design specifications of that publication series for both AMS-TeX and AMS-LaTeX.

Authors with FTP access may retrieve an author package from the Society's Internet node `e-MATH.ams.org` (130.44.1.100). For those without FTP access, the author package can be obtained free of charge by sending e-mail to pub@math.ams.org (Internet) or from the Publication Division, American Mathematical Society, P.O. Box 6248, Providence, RI 02940-6248. When requesting an author package, please specify AMS-TEX or AMS-LATEX, Macintosh or IBM (3.5) format, and the publication in which your paper will appear. Please be sure to include your complete mailing address.

Submission of electronic files. At the time of submission, the source file(s) should be sent to the Providence office (this includes any TEX source file, any graphics files, and the DVI or PostScript file).

Before sending the source file, be sure you have proofread your paper carefully. The files you send must be the EXACT files used to generate the proof copy that was accepted for publication. For all publications, authors are required to send a printed copy of their paper, which exactly matches the copy approved for publication, along with any graphics that will appear in the paper.

TEX files may be submitted by email, FTP, or on diskette. The DVI file(s) and PostScript files should be submitted only by FTP or on diskette unless they are encoded properly to submit through e-mail. (DVI files are binary and PostScript files tend to be very large.)

Files sent by electronic mail should be addressed to the Internet address pub-submit@math.ams.org. The subject line of the message should include the publication code to identify it as a Memoir. TEX source files, DVI files, and PostScript files can be transferred over the Internet by FTP to the Internet node e-math.ams.org (130.44.1.100).

Electronic graphics. Figures may be submitted to the AMS in an electronic format. The AMS recommends that graphics created electronically be saved in Encapsulated PostScript (EPS) format. This includes graphics originated via a graphics application as well as scanned photographs or other computer-generated images.

If the graphics package used does not support EPS output, the graphics file should be saved in one of the standard graphics formats—such as TIFF, PICT, GIF, etc.—rather than in an application-dependent format. Graphics files submitted in an application-dependent format are not likely to be used. No matter what method was used to produce the graphic, it is necessary to provide a paper copy to the AMS.

Authors using graphics packages for the creation of electronic art should also avoid the use of any lines thinner than 0.5 points in width. Many graphics packages allow the user to specify a "hairline" for a very thin line. Hairlines often look acceptable when proofed on a typical laser printer. However, when produced on a high-resolution laser imagesetter, hairlines become nearly invisible and will be lost entirely in the final printing process.

Screens should be set to values between 15% and 85%. Screens which fall outside of this range are too light or too dark to print correctly.

Any inquiries concerning a paper that has been accepted for publication should be sent directly to the Editorial Department, American Mathematical Society, P. O. Box 6248, Providence, RI 02940-6248.

Selected Titles in This Series

(Continued from the front of this publication)

(See the AMS catalog for earlier titles)